About Island Press

Since 1984, the nonprofit Island Press has been stimulating, shaping, and communicating the ideas that are essential for solving environmental problems worldwide. With more than 800 titles in print and some 40 new releases each year, we are the nation's leading publisher on environmental issues. We identify innovative thinkers and emerging trends in the environmental field. We work with world-renowned experts and authors to develop cross-disciplinary solutions to environmental challenges.

Island Press designs and implements coordinated book publication campaigns in order to communicate our critical messages in print, in person, and online using the latest technologies, programs, and the media. Our goal: to reach targeted audiences—scientists, policymakers, environmental advocates, the media, and concerned citizens—who can and will take action to protect the plants and animals that enrich our world, the ecosystems we need to survive, the water we drink, and the air we breathe.

Island Press gratefully acknowledges the support of its work by the Agua Fund, Inc., The Margaret A. Cargill Foundation, Betsy and Jesse Fink Foundation, The William and Flora Hewlett Foundation, The Kresge Foundation, The Forrest and Frances Lattner Foundation, The Andrew W. Mellon Foundation, The Curtis and Edith Munson Foundation, The Overbrook Foundation, The David and Lucile Packard Foundation, The Summit Foundation, Trust for Architectural Easements, The Winslow Foundation, and other generous donors.

The opinions expressed in this book are those of the author(s) and do not necessarily reflect the views of our donors.

WICKED ENVIRONMENTAL PROBLEMS

Wicked Environmental Problems

MANAGING UNCERTAINTY AND CONFLICT

Peter J. Balint, Ronald E. Stewart,
Anand Desai, and Lawrence C. Walters

Washington | Covelo | London

Copyright © 2011 Island Press

All rights reserved under International and Pan-American Copyright Conventions. No part of this book may be reproduced in any form or by any means without permission in writing from the publisher: Island Press, 1718 Connecticut Avenue NW, Suite 300, Washington, DC 20009

Island Press is a trademark of The Center for Resource Economics.

Library of Congress Cataloging-in-Publication Data

Wicked environmental problems : managing uncertainty and conflict / Peter J. Balint ... [et al].
   p. cm.
Includes bibliographical references and index.
ISBN-13: 978-1-59726-474-7 (cloth : alk. paper)
ISBN-10: 1-59726-474-1 (cloth : alk. paper)
ISBN-13: 978-1-59726-475-4 (pbk. : alk. paper)
ISBN-10: 1-59726-475-X (pbk. : alk. paper) 1. Environmental management. 2. Environmental policy. 3. Environmental protection. I. Balint, Peter J., 1950-
GE300.W53 2011
363.7–dc22
2011005206

Printed on recycled, acid-free paper ♻

Manufactured in the United States of America

10 9 8 7 6 5 4 3 2 1

Keywords: Island Press, wicked problems, wicked environmental problems, environmental conflicts, natural resource management, management of the commons, Sierra Nevada Forest Plan Amendment, Ngorongoro Conservation Area, learning networks, Comprehensive Everglades Restoration Plan, precautionary principle, adaptive management, public participatory processes

CONTENTS

Preface ix

Chapter 1	The Challenge of Wicked Problems	1
Chapter 2	Risk and Uncertainty in Environmental Management	7
Chapter 3	Four Wicked Cases	33
Chapter 4	The Precautionary Principle	65
Chapter 5	Adaptive Management	79
Chapter 6	Participatory Processes	103
Chapter 7	A Proposed Adaptive, Deliberative Decision Process	129
Chapter 8	The Sierra Nevada Example: Survey of Stakeholders	149
Chapter 9	The Sierra Nevada Example: Elicitation and Analysis of Preferences	167
Chapter 10	Managing Wicked Environmental Problems	207

References 219

Authors 237

Index 239

PREFACE

This book is based on our work with the United States Department of Agriculture Forest Service (hereafter referred to as the Forest Service) and its efforts to amend national forest plans in the Sierra Nevada region of California. During our research, we came to the conclusion that this decision dilemma meets the requirements of a *wicked* problem. Wicked problems are characterized by a high degree of scientific uncertainty and a profound lack of agreement on values. Further, even though there is no correct decision in the case of a wicked problem, the manager must make a decision. The identification of the Sierra Nevada planning effort as a wicked problem leads to a critical conclusion. Because, by definition, a wicked problem has no optimal solution, the decision maker must seek other measures of success. The book traces our research and findings and proposes an approach to managing or coping with such problems.

Our work began in 2003 when Jack Blackwell, the regional forester for the Pacific Southwest Region of the Forest Service, asked Ronald Stewart to put together a team to answer the question, "How did the region deal with risk and uncertainty in the Sierra Nevada Forest Plan Amendment final environmental impact statement and record of decision signed in January 2001?" The research team consisted of four people with diverse qualifications. Ron Stewart's background is in forest ecology and Forest Service administration. As regional forester in the Pacific Southwest region from 1990 to 1994, he initiated the Sierra Nevada Forest Plan Amendment process when it began in 1992. Peter Balint's experience is in conservation biology and environmental policy. Larry Walters, who once served as a town supervisor, is an expert in public finance and public administration. Finally, Anand Desai brings expertise in public policy analysis and modeling. The team's diversity of training in theoretical, practical, and analytical approaches in both natural and social sciences, combined with personal experience in the specific context of the Sierra Nevada decision dilemma, led us to explore and integrate ideas from a wide range of disciplines.

As we examined the Sierra Nevada case, and three other similar domestic and international environmental planning efforts, we concluded that environmental management agencies rarely identify a problem as wicked even after repeated failed attempts to reach a satisfactory conclusion. Instead, management decisions are typically followed by unproductive cycles of appeals and litigation, failed implementation, and new rounds of analysis and public participation. Each round may include more sophisticated analysis, greater public engagement, and longer and more complex documents, but it inevitably leads to the same conflicted outcome. This failed approach assumes that reducing scientific uncertainty and improving public understanding of the problem will lead to a solution. Our research, however, led us to believe that, while arguments in the context of a wicked problem may be framed around science and scientific uncertainty, the real issue is often deep disagreement on values. In a wicked problem, key stakeholders, including the agency and various interest groups, typically have significantly different and often incompatible worldviews. Yet these profound differences are rarely acknowledged or explored. Thus a missing dimension in the decision process is an effort to explicitly identify and consider the range of values that inform participants' perceptions of the problem and their preferred policy responses.

The defining characteristics of a wicked problem—a high degree of scientific uncertainty and a profound lack of agreement on values, combined with the absence of a perfect solution—led us to propose an approach that builds on the idea of *learning networks*. In a learning network, participants engage in an iterative, analytic, deliberative process to build trust and move toward agreement. In our research, we tested novel techniques to identify public and agency values and preferences and incorporate them into ecological models. The outcomes of these combined models can be used to develop alternative management choices that may otherwise be overlooked but may have the potential to attract broad support. We suggest that the information generated through these techniques could serve as new input in the learning network to help participants move forward. We further recommend that any decision that emerges should be implemented using an adaptive management philosophy to allow flexibility in adjusting to the complexities and uncertainties inherent in wicked problems.

We gratefully acknowledge the financial support of the USDA Forest Service Pacific Southwest Region for the study of the Sierra Nevada Forest Plan Amendment process that formed the basis for this book. We especially recognize the contribution of time and ideas from Regional Forester Jack Blackwell, Deputy Regional Forester Kent Connaughton, Sierra Nevada Forest Plan Amendment Review Team Leaders Mike Ash and Kathy Clement, and Public Affairs Officer Rick Alexander.

We thank the participants in three workshops we conducted in the region. These included members of the Forest Service Regional Management Team and stakeholders from the general public who traveled to meetings in Sacramento to participate in discussions with us and complete two data collection exercises that helped us understand the Forest Service's planning problem.

We acknowledge the help of research assistants Beena Chundevalel, Nancy Kanbar, Melissa Milne, and David Phillips at George Mason University, and Yija Jing at The Ohio State University. We appreciate the work of the editors at Brigham Young University who reviewed early drafts of our chapters.

We are grateful to Barbara Dean and Erin Johnson at Island Press for their editorial guidance. We thank Marce Rackstraw for creating the graphics.

Chapter 1

The Challenge of Wicked Problems

For almost a century, advocates for preservation and for development have argued about the effects of human actions on the environment. These arguments have been made more difficult to resolve because there are still considerable uncertainties in science, and because it takes a long time for the effects of human actions to show up in the environment. Both sides, and other groups who fall along a continuum between them, have exploited these uncertainties in appeals and litigation. The logical result was for government agencies to produce more complex documents justifying their decisions and to include and advocate for more science, causing many to assume these disputes were based in science. But we believe the evidence shows that the underlying differences in stakeholder positions are not so much related to uncertainties in science or failure to consider particular aspects of the scientific literature, but rather to conflicting values and preferences, and therefore differing views on desirable outcomes. These elements of the argument are rarely, if ever, considered in the decision-making process. As a result, most environmental arguments continue to produce more detailed documents and longer processes without resolving the underlying issues.

Wicked Problems

The clashing interests of environmentalists, developers, and others have elevated many environmental problems that require decisions at the federal and state level from simple, to complex, to "wicked" (Salwasser 2004; Lackey 2007). A wicked problem is characterized by a high degree of scientific uncertainty and deep disagreement on values (Allen and Gould 1986; Committee of Scientists 1999). The definition of a wicked environmental problem itself is in the eye of the beholder, or the stakeholder, and therefore there is no single correct formulation of any particular problem (Rittel and Webber 1973; Allen and Gould 1986). Consequently, there is no single, correct, optimal solution. The decision maker must come to a conclusion without knowing if all feasible and desirable options have been explored, and any management choice will ultimately be better or worse rather than true or false.

In this book, we examine the class of wicked problems, including proper identification of such problems and how they have been dealt with in the past. We propose a modified decision-making approach that blends current thinking on addressing wicked problems and stakeholder participation with our understanding of the best practices already implemented by agencies to address such problems. Our approach relies on developing a *learning network* among the stakeholders, using an *adaptive, iterative, deliberative, analytical participatory process*. An important component of this method is incorporating stakeholder preferences into the ecological models that resource management agencies currently use to support decision making. We also suggest that since wicked problems have no single best solution, decision makers must seek management policies and processes that are "satisficing"—that is, potentially broadly acceptable and implementable—rather than optimal. Herbert Simon (1957) coined the term satisfice, combining the words "satisfy" and "suffice." A satisficing strategy accepts an outcome or judgment as good enough or satisfactory without an expectation that it is in any sense optimal or best.

In this book we also touch on the important consequences of properly or improperly identifying a wicked problem. Not all problems rise to the level of wicked, but when they do, the processes used become critical. Although environmental dilemmas may occasionally meet the criteria for wicked problems, such problems are by no means confined to the environment. Whenever interest groups with strongly divergent values are well organized and highly motivated, and uncertainties in the science may be exploited, an issue can move into the realm of a wicked problem.

Historical Perspective of Environmental Controversy

In this book, we introduce and discuss case studies of wicked environmental problems in the United States, Europe, and Africa. Our key case study, however, focuses on national forest management in the Sierra Nevada region of California. In presenting a brief summary of political conflicts over the environment in this section, we emphasize the origins of these disputes in the context of public lands in the western United States. While the details of the social and historical trends leading to environmental conflict differ across our case studies, there are also, as we discuss in the book, significant common factors, including, most importantly, scientific uncertainty and profound differences in perceptions, attitudes, and values among key stakeholders.

From today's perspective, many view the age of environmental controversy as beginning in the 1960s. However, the battle over environmental management among prodevelopment, propreservation, and other interest groups in the United States has a history more than a century long. Unfortunately, the resulting political compromises have not addressed the fundamental and underlying differences in public values represented by these positions. Because these value differences were not taken into consideration, stakeholders have continued to press their arguments through the courts using the laws passed in the 1960s and 1970s that opened the federal and state decision-making processes to public participation.

The US environmental movement had its roots in battles over the public domain in the western part of the country. The initial philosophy of Congress and the federal government during the mid- to late 1800s was to encourage settlement and development by disposal of these lands to railroads, farmers, and others. Various acts of Congress encouraged mining and oil production to meet the needs of a growing population and economy. However, a significant change in attitude toward the remaining public lands began to emerge in the last quarter of the nineteenth century. Publication of the *Report Upon Forestry* (Hough 1878) and meetings of the American Forestry Association demonstrated a growing concern about the overharvesting of forests and overgrazing of public lands. The establishment of Yellowstone, the first national park, on March 1, 1872, and passage of the Forest Reserve Act of 1891 signaled the end of disposal to private interests and the beginning of federal ownership for protection of natural resources. The Forest Reserve Act allowed the president of the United States to designate forest reserves. Lands so designated were protected from disposal, but the act did not include any administrative authority for the use of those lands (Steen 1992). However, concerns about loss of development potential for local communities resulted in pressure by western members of Congress to allow timber harvesting and grazing, leading in turn to passage of the Organic Act of 1897. The Organic Act provided for watershed protection and included an implied goal of long-term sustainability for the nation's natural resources. This compromise earned initial support for setting aside additional forest reserves among members of Congress from both the western and eastern areas of the country.

This compromise was short-lived, however. In 1907, western interests moved to block the president's authority to establish forest reserves through the annual agricultural appropriation bill (Steen 1992). The catalysts for this action were a series of land fraud trials in Oregon and President Theodore Roosevelt's aggressive establishment of new reserves. With this act, the authority to establish additional reserves resided exclusively with Congress. Roosevelt, with the help of the first chief of the US Forest Service, Gifford Pinchot, moved quickly to establish an additional sixteen million acres of reserves before the law took effect (Steen 1992). The bill also changed the name from "forest reserves" to "national forests."

With the rise in power of the organized environmental movement in the 1960s and the passage of both state and federal legislation that opened up the decision-making process to public review and gave citizens the right to sue the government, the site for environmental battles came to include the courts as well as Congress and the state legislatures. The National Environmental Policy Act of 1969, Endangered Species Act of 1973, Clean Air Act of 1970, Clean Water Act of 1972, and other statutes required that federal decisions affecting the environment must be open to public involvement. Provisions for citizens and environmental groups to litigate over decisions that they did not support were also provided. Most states followed suit with similar laws and regulations.

These environmental laws also required that decisions be accompanied by detailed comprehensive analyses of alternatives and their potential impacts (environmental impact statements) and written documentation of the decision and its justification (the formal record of decision or finding of no significant impact). All these documents were subject to public review and comment, and the agencies were required to explain how they responded to that input. Because government agencies are delegated the responsibility to make decisions by law, opponents of a decision must base their legal arguments on procedural deficiencies or failures to comply with specific requirements of law or agency regulations. Procedural deficiencies may include inadequate public involvement, failure to adequately consider other alternatives, failure to adequately consider public input, or failure to consider or properly interpret science. Successful litigation on these issues has prompted agencies to create lengthier, more complex, and more analytical documents in an attempt to address these potential grounds for lawsuits. This in turn has resulted in more protracted and involved public participation processes. Since this process ultimately does not address the fundamental underlying issues—disagreement over values and dissatisfaction with the decision itself—it often results in continuing litigation and in a cycle of decisions that cannot be implemented.

Overview of the Book

Here we briefly summarize the focus of each of the remaining chapters. Chapter 2 formally introduces the concept of wicked problems. In this chapter we describe the characteristics of wicked problems, and discuss the ways uncertainty, risk, divergent values, and other factors contribute to the *wickedness* of these problems.

Chapter 3 presents the four case studies that serve as examples throughout the book. The first three sections of the chapter examine problems associated with efforts to restore the Everglades in Florida, manage the Ngorongoro Conservation Area in Tanzania, and implement a cap-and-trade program for carbon dioxide emissions in the European Union. The fourth section introduces our primary case study—the problems associated with managing the Sierra Nevada national forests of California. Through these diverse cases, we illustrate that the concept of wicked problems has broad applicability across a variety of natural resource management dilemmas in both developed and developing countries. The discussion of the four cases highlights both the characteristics that these dilemmas have in common and the challenging idiosyncrasies that make them resistant to generalized policy responses.

Chapters 4 through 6 examine ways managers have commonly attempted to address these kinds of complex dilemmas, whether or not they explicitly understood that they were facing a wicked problem. Chapter 4 focuses on the precautionary principle, which advocates proactive efforts to anticipate and reduce the likelihood of future harms. Chapter 5 discusses adaptive management, which incorporates an acceptance of limits to current knowledge and applies systematic efforts to promote learning from carefully designed and monitored management experiments. In these chapters, we also consider ways in which the precautionary principle and adaptive management may conflict with each other. In chapter 6, we describe the role of public participation in managing complex environmental problems. This approach, an essential component of modern democratic processes, is now widely required by law. It also provides a clear avenue for the expression and inclusion of diverse public values in the policy process. In our discussion, we also consider common challenges that may limit the effectiveness and efficiency of participatory processes.

In chapter 7, we recommend an approach designed to incorporate and improve on the decision principles and processes used to date in the context of wicked problems. Our recommended approach builds on the learning network process proposed in the literature (National Research Council 1996), incorporates the procedural requirements of the National Environmental Policy Act, and adds novel methods for formally eliciting and analyzing public values.

In chapters 8 and 9, we describe the results of a pilot test of our proposed analytic methods in the context of the Sierra Nevada case study. During our research to gather information on stakeholder attitudes and preferences, we held three workshops in the region. During these meetings, we asked participants to complete a questionnaire on their perceptions of the decision process and a card-sort exercise in which they could consider and rank various policy alternatives. In chapter 8 we report the results of the questionnaire, and in chapter 9 we describe our analysis and findings from the card-sort exercise.

Finally, in chapter 10 we summarize our views on how decision makers and managers might best cope with wicked problems. While acknowledging that not all problems are wicked, we emphasize that the appropriate identification of a problem as wicked can itself be useful for the public manager. This identification has important consequences. For example, since a wicked problem has no optimal solution, the manager—while still required to act—is released from the impossible task of finding the one *correct* response. Given the idiosyncratic diversity and apparent intractability of wicked problems, we do not claim that our approach can transform wicked problems into tame ones or that it will fit all circumstances as a fixed template. But we believe our proposal has the potential to facilitate progress and may usefully be adapted to match the varying contexts of wicked problems.

Chapter 2

Risk and Uncertainty in Environmental Management

When it comes to environmental conflict, what makes some decisions more difficult than others? For example, the state of California routinely experiences thousands of wildfires each year, hundreds of which are the natural result of lightning strikes. If these naturally occurring phenomena are so common, what makes decisions related to the management of these situations so challenging?

Similarly, there had been an apparent consensus regarding development strategies in the Everglades in Florida; however, as regular flooding and polluted streams indicate, those strategies are not sustainable. And yet, there appears to be no agreement between those who favor preservation and those in favor of development on alternative solutions. How do situations that are used to derive a consensus on how to address the situation suddenly become a source of contention?

Such intractable problems are not unique to the United States. The European Union has been able to make little headway toward implementing market strategies to reduce carbon dioxide emissions that seem to have worked, at least partially, in the United States. Likewise, in Tanzania government agencies have yet to create a clear path to balancing the competing needs of a homeland for the Maasai, wildlife preservation, and tourism revenues.

To appreciate how these and similar worldwide decisions differ qualitatively, we must first review traditional approaches to problem structuring and decision making, particularly in the analysis of public problems. This process of evaluating and choosing from alternatives is often iterative, but there are relatively well-defined, sequential steps that analysts employ in developing an effective public policy (Kweit and Kweit 1987; Patton and Sawicki 1993; Dunn 1994; Bardach 1996; McRae and Whittington 1997; Walters, Aydelotte, and Miller 2000).

1. Define the problem.
2. Identify the criteria to be used in evaluating alternative solutions.
3. Generate alternative solutions to the problem.
4. Evaluate the alternative solutions based on the evaluation criteria.
5. Recommend an alternative.

Even practical approaches to improved individual decision making often parallel these steps (Hammond, Keeney, and Raiffa 1999). How well this general approach will work depends in part on the nature of the issue at hand. Several authors have pointed out that the structure of public problems can be characterized along several dimensions. Walters, Aydelotte, and Miller (2000) offer the following list of factors to help predict how serious a given problem is.

- The degree of conflict over the issue
- The number of stakeholders
- The level of confidence in the information on the issue
- The number of alternatives
- The knowledge of outcomes
- The probability of the outcomes

The result is a continuum beginning with well-structured problems at one pole and "ill-structured" problems at the other (Mitroff and Sagasti 1973; Dunn 1994; Walters, Aydelotte, and Miller 2000). Many of society's pressing problems—in environmental management and elsewhere—possess high levels of complexity and social conflict, as well as profound social and cultural values incompatibility. In these most complex cases, the processes of defining a society's shared values, common goals, desirable outcomes, and acceptable risks become political. In such cases, the generally accepted approaches to problem structuring and analysis crumble, and consequently, technical analyses alone—which do not integrate social values and deliberation—cannot provide an adequate decision-making framework. In other words, when scientific uncertainty coexists with value uncertainty and conflict, we have wicked problems.

Defining Wicked Problems

When government officials and citizens make decisions in the public arena, the decisions occur at various levels of complexity. Some decisions are difficult to analyze, understand, or explain. Problems take on a complexity that often extends well beyond the merely intricate and assumes many forms, including high levels of risk; scientific uncertainty; biological complexity; social complexity; vast scope and scale of the issues involved; and the absence of a clear public consensus on values, the nature of the problem, or acceptable solutions.

Clearly, some public problems are more difficult to resolve than others. Renn (1995) suggests that environmental debates operate on three levels and that ecological risk assessment is less and less helpful in policy making as levels of complexity and conflict increase. For straightforward (well-structured) problems, scientific analysis and traditional analytic approaches may serve as a basis for policy making with little controversy. At a medium level of complexity, public trust in the implementing institutions and their technical expertise is required. It is at the highest level of complexity and conflict that political forces overshadow technical analyses, making stakeholder involvement absolutely essential.

Paris and Reynolds (1983) observe that policy decisions inevitably involve three claims:

- empirical claims about causal relationships, observable levels of key variables, and generally (potentially) verifiable statements about the world;
- normative claims that focus attention on particular concerns and judge the acceptability of the status quo, outcome levels, or important relationships;
- action claims that assert the need for particular policy changes consistent with empirical understandings and in light of normative judgments.

Consider two dimensions of any decision: the state of empirical knowledge necessary to make the decisions, and the level of agreement on guiding values (see table 2.1).

TABLE 2.1. Decision problems, from easy to wicked

	Agreement on Values	
State of knowledge	High	Low
Well developed	Routine analysis with periodic stakeholder and expert review. Decisions are easy.	Emphasis on stakeholder deliberation with periodic expert review.
Tentative/gaps/ disagreements/ research needed	Emphasis on expert deliberation with periodic stakeholder review.	Emphasis on both stakeholder and expert deliberation. Wicked problems!

Given these two dimensions, there are four possible scenarios:

1. If decision makers understand and generally accept the knowledge base underpinning an issue, as well as agree on what values are most important, then decision making is relatively straightforward and stakeholders may be comfortable with a strategy proposed by an agency or expert.
2. If decision makers do not agree on values, but the science is well understood, then the focus is on dialogue among the stakeholders to understand and resolve value differences.
3. When the science is uncertain and there are important gaps in the knowledge base, but stakeholder value agreement is high, then the focus is on resolving the science issues with oversight and, when needed, with the stakeholders' assurance that their values are reflected in the science and decision making.
4. But when both the science is uncertain and value agreements are low, then the issue will likely become a wicked problem and need significant and repeated dialogue among scientists, stakeholders, and decision makers.

Raiffa (1968) defines a decision problem as a choice among a set of actions. The ideal decision entails selecting the action that optimizes the decision maker's return, where each outcome is assigned a worth or utility to the decision maker. Hence an outcome is associated with an action that is taken in a given context. When the relationship between an action and its outcome is clear, we have a programmed decision (Simon 1960, 5), which falls into the top left-hand quadrant of table 2.1. Problems belong to the top right-hand quadrant when there is uncertainty about goals, values, and objectives, and consequently the utility associated with that action is unclear. This ambiguity is not due to any confusion on the part of the decision maker regarding her personal priorities; instead, it is because she must act on behalf of the public at large. But in the case of the right column of table 2.1, there is no agreement among publics about the values and goals they want to pursue. Some will label whatever the decision maker elects to do as nonresponsive. Problems in such contexts are generally addressed through political means, and the decision makers arrive at the solutions through debate and compromise. On the other hand, when values are clear and the utilities certain, but the outcome corresponding to the action in a given context is not known with certainty, additional information is needed. Problems whose solutions have uncertain and potentially unknown consequences belong in the bottom left-hand quadrant. An unprogrammed decision (Simon 1960) or wicked problem (Churchman 1967; Rittel and Webber 1973) has one or more of the actions, the context, the outcome, or the utility totally unknown or not confidently known (Mason and Mitroff 1973).

Setting Wicked Problems Apart from the Rest

The nature of wicked problems is such that it is difficult to generalize about them; however, they seem to have a number of common characteristics. For example, selecting a solution from a set of limited options solves the usual decision problem. These options are well defined with stable problem statements such that one knows when a solution is reached. Hence, it is possible to objectively evaluate the solutions as either right or wrong, to try out the solutions and abandon them if they do not work (Conklin 2006).

In sharp contrast, the definition of a given wicked problem is in the eye of the beholder; that is, each stakeholder defines the problem differently and therefore there is no uniquely correct formulation of the problem. Because a number of factors, such as resources of ecosystems, communities of interest, funds, and organizational capabilities, combine with stakeholder demands in idiosyncratic ways, any resolution is likely to be one-shot and unique. Also, outcomes are not scientifically predictable, so the decision maker cannot know when researchers have explored all feasible and desirable solutions. In fact, responses to wicked problems are generally better or worse, rather than right or wrong, and it may take a long time before the real consequences of a decision are discovered (Allen and Gould 1986).

These characteristics result in some disturbing problem attributes. Here are the ten propositions offered by Rittel and Webber (1973, 162–67) as distinguishing properties and outcomes of wicked problems:

1. There is no definitive formulation of a wicked problem.
2. Wicked problems have no stopping rule.
3. Solutions to wicked problems are not *true* or *false*, but *good* or *bad* or *better* or *worse* or *satisfying* or *good enough*.
4. There is no immediate and no ultimate test of a solution to a wicked problem.
5. In a wicked problem, there is no opportunity to learn by trial and error. Every solution is a one-shot operation.
6. Wicked problems do not have an enumerable (or an exhaustively describable) set of potential solutions, nor is there a well-described set of permissible operations that may be incorporated into the plan.
7. Every wicked problem is essentially unique.
8. Every wicked problem can be considered a symptom of another problem.
9. The existence of a discrepancy representing a wicked problem can be explained in numerous ways. The choice of explanation determines the nature of the problem's resolution.
10. The planner has no right to be wrong.

These propositions focus primarily on two aspects of the problem: its definition and the nature of the solution. As propositions 1 and 7 suggest, attempting to formulate the problem is itself a problem. Further, because of the situation's uniqueness, it is not always possible to turn to other similar situations for potential insights. In a wicked problem, there is ambiguity about the nature of the problem. Often there is no single problem but a combination of multiple intractable problems that are unearthed during the process of problem definition. If we think of a problem as a discrepancy between the current state of affairs and a desired state (proposition 9), then the solution has to eliminate the discrepancy. Hence, how we choose to explain the discrepancy will determine the type of solution we seek.

However, there are no criteria that tell when *the* solution or *a* solution has been found in dealing with wicked problems. Because (a) the process of solving the problem is identical to the process of understanding its nature (proposition 9), (b) there are no criteria for determining what is a sufficient understanding of the underlying issues (propositions 4 and 6), and (c) there are no ends to the causal chains that link interacting open systems—the manager/planner can always invest more efforts to increase the chances of finding a better solution (proposition 2). Rather than solving it, the manager often terminates work on a wicked problem for external considerations: not enough time, money, or patience.

Rittel and Webber (1973) suggest that even short-term "solutions" do not end wicked problems (proposition 4) because the problems are dynamic, and social and scientific parameters will change over time. With wicked problems, any solution implemented will generate waves of consequences over an extended period of time. Additionally, there is no way of tracing these waves through all the affected lives since the full consequences cannot be appraised until the waves of repercussions have completely run their course—which may, in the case of issues involving forest ecosystems, take decades or even centuries.

Decision makers disagree on the exact definition of any particular wicked problem; consequently, the criteria are not clear for judging solutions. Judgments regarding whether a solution is true or false are likely to differ widely depending upon the stakeholder community or personal interests and values (proposition 3).

Living with Wicked Consequences

Although one can learn lessons from implementing solutions, proposition 5 raises an interesting issue about the utility of the lessons learned for the current problem. In saying that "there is no opportunity to learn by trial and error," Rittel and Webber (1973) are not suggesting that there are no lessons to be learned, but rather that the lessons learned will come too late to help with the problem at hand. By this time, the situation has evolved into something different (proposition 8), which requires a redefinition and reformulation of the situation that now needs addressing. To illustrate, Rittel and Webber give the example of building a freeway where the implementation of the decision has long-term consequences and is not readily reversible. So it becomes important that there be a general consensus regarding the course of action and a willingness to live with the consequences.

Traditional decision theory (Raiffa 1968) focuses on the selection of an option from a set of differently desirable choices, each of which has its own costs and benefits. Proposition 6 however, suggests that such a set of potential solutions does not exist for wicked problems, in part because there are no criteria that enable someone to prove that all relevant solutions have been considered. With these ill-defined problems and solutions, the set of feasible plans of action relies on realistic judgment and on the amount of trust and credibility between policy makers and the public, which may be small or nonexistent.

And finally, proposition 10 draws the distinction between an administrator's and a scientist's job. In science, solutions to problems are considered hypotheses to be refuted. And, the scientific community does not blame its members for postulating hypotheses that are later refuted. In dealing with policy issues as they relate to wicked problems, however, planners are liable for the consequences of their actions or inactions. Here, the aim is to find ways to improve some characteristics of our world; thus a policy's effects can matter a great deal to people touched by the actions taken.

Natural Resource Problems

Most, but not all, large-scale planning issues involving the *commons* have become controversial. Public values combine with issues of scientific uncertainty and geographic scale to create wicked problems. Citizens are concerned about public lands, oceans, and the atmosphere meeting natural resource supplies, accommodating rural community demographic changes, and adjusting to declining populations of certain plants and animals. Salwasser (2002) addresses the nature of natural resource problems in today's decision environment, characterizing them by

- their complexity and messiness: no definitive problem statement, and multiple problems with multiple objectives;
- the existence of fragmented stakeholders: both in interests and in tactics used to pursue their interests;
- scientific messiness: multiple factors influence each problem area or objective, and the manager can only influence some of these factors;
- two kinds of uncertainty: (1) we do not know but can eventually learn, and (2) we cannot know until it occurs; to this we add a third—we do not know that we do not know;
- conflicting risks: there are conflicting risks among objectives and between short-term and long-term objectives; and
- dynamic social, economic, knowledge, and technological systems.

The sociopolitical and environmental systems involved in natural resource issues have both time and spatial dimensions. For instance, the regulations implementing the National Forest Management Act of 1976 require that the US Forest Service maintain viable species populations throughout their range when considering forest plans and individual projects. Environmental activists are concerned that a piecemeal approach with individual plans made in a portion of a species' range might cumulatively jeopardize the species' long-term viability as each separate individual plan is implemented. The worry is that the cross- or multijurisdictional nature of species (which do not adhere to human-made boundaries) has increasingly forced the agency to consider creating large, landscape-scale planning efforts to prevent cumulative negative effects of incremental decision making. At the same time, the planning regulations require preparing and implementing forest plans by each local planning unit, generally the individual national forest. Planning across distinct administrative units increases complexity and adds to the number of issues and problems that must be addressed in the planning process, including the number of stakeholder participants. The complexity is further compounded when decision makers have to consider transnational issues, for instance, in the European Union.

In addition to these scale issues, the long time frames of ecological response, and the short time frame of the sociopolitical process and changing societal values, result in an even more complex wicked problem. On the time scale, the sociopolitical environment can be rather volatile, changing with the next election or lawsuit. Planning is also limited in terms of human lifetimes of those people involved on the project, thus it is difficult to successfully complete multigenerational projections. On the other hand, natural environments—for example, river basins or forests—change continuously, but slowly, on the order of decades and even centuries. Experimentation to resolve issues of uncertainty may take decades; in the meantime, the sociopolitical process may demand faster resolution or change in direction before we can know if the old direction was satisfactory. This is further complicated by the long time frame of a complex, large-scale planning process, in which the issues that initially framed the planning process could change before the decision is reached.

At the landscape scale, there are issues of both local and universal concern, especially with respect to species that have wide ranges. Further, all issues have both a stakeholder community of place—those most immediate to the affected area—and a stakeholder community of interest—a broader group that can live anywhere, and who, if disenfranchised, can resort to competitive strategies to assert their rights. Those interested in community-of-place issues can often find more ready agreement because they share a common interest in the local community and have to coexist after the debate ends and the decision is made. For these same reasons, it is much harder for the broader community of interest to make necessary compromises and move forward. It is also more difficult for the decision makers to identify future stakeholders and to find ways to meaningfully engage them in the planning process. Yet, given the controversy over managing the nation's forests, wetlands, climate, and so forth, there are likely to be few issues that are purely of local community interest, assuming there exists such a thing as a static local community.

Thus, wicked problems are extremely complex and generally unsolvable. However, and perhaps because of their complexity and seeming intractability, there is a growing body of literature and practical experience contributing to understanding and, perhaps, *managing* such problems.

Understanding Open and Closed Ecosystems

In the past several decades, there has been a major change in understanding ecosystems. Formerly, the dominant paradigm was of ecosystems that, when mature, were stable. They were thought to be closed, unaffected by external influences, deterministic, and self-regulated. If this stable condition were to be disturbed, an ecosystem was expected to progress through a series of successive stages back to its original, stable, homeostatic state (Daly 1993). Nature, unfortunately, does not work this way.

The current paradigm is that open systems are in constant states of flux, affected by a series of stochastic factors originating both inside and outside the ecosystems. As a result, these systems are probabilistic and multicausal rather than deterministic and homeostatic like closed systems (Daly 1993). The current model also recognizes that human impacts almost always play an important, and often dominant, role in affecting a system's status (Smith 1997). Present knowledge also emphasizes that uncertainty is central to managing living resources. It follows, then, that ecosystems are characterized by uncertainty—in their basic ecology and biology, in their economic parameters, in the effect(s) of management actions, and there is even uncertainty as to whether or not it is possible to achieve management objectives. Therefore, policy makers, managers, and the public must recognize uncertainty as an overriding factor.

Diversity of Values

Adding to the complexity of environmental management decisions is the range and diversity of public stakeholders' values. We define values as concepts or beliefs about desirable behaviors or states that transcend specific situations, guide the selection or evaluation of behaviors and events, and are ordered by relative importance (Schwartz and Bilsky 1987). One key aspect of values is that they rationalize actions (Rescher 1969). Whether the issue is protecting old-growth forest species, wetland habitats, air quality, or aquifers, there is little doubt that stakeholders differ in their fundamental values, their conceptualization of environmental issues, and their priorities for the future. Decision makers must recognize this diversity of views, but strategies for managing value conflicts are often slow, cumbersome, expensive, and uncertain in their own right.

Managing in the face of uncertainty has an added essential requirement: to identify and characterize risk. The presence of value conflicts, risk, and uncertainty does not mean that a definite management decision cannot be made, but it emphasizes the manager's need to think in terms beyond the traditional approach to problem structuring and problem solving. When considering the future and the consequences of current managerial actions, it is essential to identify distributions of likely futures and a consequent range of effects of subsequent management efforts rather than single-point estimates. Rigorous modeling is required to define the safe or precautionary approach: modeling to simulate the whole system under management, applying sensitivity analysis to determine the stochastic elements in the system, and finally defining through the models the likely outcomes of various management options.

Before discussing managerial options, however, we recognize that ecosystem *uncertainty* and *risk* are concepts that must be defined and distinguished. Further, we need to clarify the concept of values and develop a framework for characterizing values. The treatment of these concepts varies across the social sciences. Decision-making processes underlying many environmental issues seem to draw upon the various conceptualizations found in the psychology, anthropology, sociology, geography, and economics literatures. We summarize the main themes from these disciplines to provide insights into the meanings associated with these different notions of uncertainty, values, and risk as they are used in deliberations on managing complex environmental settings.

Uncertainty

Given the nature of science and its development, scientific *truths* are always subject to review and revision. Scientific certainty is a probabilistic notion; hence, it is exceedingly rare for a large group of scientists to agree with certainty about anything, especially about something as complex as an environmental or ecosystem-level problem. When talking about living systems, great scientific uncertainty is the norm: in our complex dynamic environment, knowledge has limits and certainty is difficult to attain. Thus, uncertainty far outweighs knowledge of cause and effect.

Categories of Uncertainty

Generally, uncertainties can be placed in the following categories (Tickner 1999):

- *Parameter uncertainty* refers to missing or ambiguous information about factors underlying uncertainties. This type of uncertainty can potentially be reduced by gathering more information or by using better collection and analysis techniques. However, if it is due to variability in these factors' natures, it may not be possible to reduce this type of uncertainty without obtaining a better understanding of the uncertainty's root causes. In attempting to determine the consequences of environmental releases, in particular, uncertainty regarding the full range of individual reactions could make it difficult to determine an individual's potential susceptibility to harm.
- *Model uncertainty* refers to gaps in scientific theory, or imprecision in the models used to bridge information gaps. As models are abstract, constructed to explain current or past events, or to predict the future, they are only as good as the information used to build them (garbage in = > garbage out). Comprehensive models of open and interdependent environmental systems are, by definition, incomplete—and therefore limited in their descriptive or predictive powers.
- *Systemic uncertainty* refers to the unknown effects of cumulative, multiple, simultaneous, and interactive exposures. Systemic uncertainty can be an important confounding factor in large-scale and/or long-term analyses.

- *Politically induced uncertainty* refers to deliberate ignorance on the part of agencies charged with protecting the environment. An agency may decide not to study a hazard, limit the scope of its analysis, and downplay uncertainty in its decisions.
- *Indeterminacy* means that the uncertainties involved are of such magnitude and variety that they may never be significantly reduced.
- *Ignorance* has two faces. Positively, it is a humble admission that we do not know how much we do not know. Negatively, it is the practice of making decisions without considering uncertainties.

Uncertainty Is Value-Free

We define uncertainty as a neutral analytical property of an event, relationship, phenomenon, or other important consideration that may be reduced through better science, but generally cannot be eliminated. In this context, uncertainty is the likelihood of the occurrence of an event, relationship, phenomenon, or other important consideration. This likelihood of occurrence may be unknown, or may have a distribution of possible values, but it is not under the immediate control of decision makers (Knight 1921).

In describing uncertainty as value neutral, we highlight two important points. First, uncertainty is used to describe probabilistic events, whether or not it is possible to quantify those probabilities. For example, if we are able to calculate the distribution of natural disasters, we may also be able to estimate the probability of different consequences during a specific time interval. But it still may be impossible to estimate the likelihood of important budget changes resulting from shifts in national public policy priorities during the next fifty years. In both cases, however, *uncertain* is the analytical term used to describe the events.

Second, uncertainty does not inherently involve a value position on the part of the analyst or decision maker. The probability of a forest fire, for example, is independent of attitudes toward fire hazard, economic development, or any other value position. In this sense, uncertainty is a neutral concept. As we discuss in the following section, values enter the discussion when considering the perceived positive or negative effects of the uncertain events, should they occur.

There are three broad categories of uncertainty in the environmental decision context: *scientific*, *administrative* (or *implementation*), and *stochastic*. To say that something is *scientifically uncertain* within the context of an environmental decision problem is to acknowledge that ecosystems are complex and that our knowledge of them is incomplete. As a result, no one can state with certainty the long-term outcomes of any given management strategy, including maintaining the status quo. Scientific uncertainty is often expressed as a calculated or estimated confidence interval around a predicted value or outcome. However, in complex systems, estimations of the likelihood of extreme outcomes as combinations of independent events often lead to severely wrong predictions. *Administrative* or *implementation uncertainty* refers to the vagaries of managing in a political environment in which public goals and priorities, societal needs and conditions, and organizational capacities change over time. Finally, *stochastic uncertainty* refers to those events that are largely random, unpredictable, and uncontrollable, such as lightning-caused ignitions or random changes in species populations.

Values

As stated previously, values rationalize action and can be defined as concepts or beliefs about desirable states or behaviors that transcend specific situations, guide the selection or evaluation of behaviors and events, and are ordered by relative importance (Rescher 1969; Schwartz and Bilsky 1987). It is worth noting that problems have no objective existence in the world. In the normal course of living, people come to expect the world to work in particular ways, to provide certain predictable experiences, opportunities, and relationships. When our expectations are not met, when the expectation is deemed sufficiently important and the resulting dissonance exceeds a certain threshold, we perceive a problem. When multiple people see the same situation as problematic, though often in different ways, we have created a social problem. Thus, three normative conditions must be present. First, a valued expectation must be unsatisfied. Second, the dissonance must exceed some level of acceptability. And third, the unmet expectation must be sufficiently important to warrant action. In the case of social problems, if there is agreement among stakeholders on the unmet expectation, the relative importance of that expectation, and the threshold of dissonance, then we say there is normative agreement. In the case of wicked problems, there is nearly always disagreement on which expectations are not being met, on the relative importance of those expectations, and the appropriate thresholds of acceptability.

While it is true that all policy decisions have a normative component (Paris and Reynolds 1993), values have come to play an even more important role in environmental decision making in recent decades. Precisely because of the uncertainty surrounding decisions that affect the environment, the public has become more interested and engaged in environmental management (O'Brien 2003). Increased public participation inherently means that decision makers and managers must attend to a wider range of values than previously. Whether public participation is viewed as a cynical method for diffusing public concerns or more constructively as a serious engagement with legitimate stakeholders, it is clear that public values vary widely, are often contested, and can change over time. Within this milieu, decision makers are struggling to make sense of the spectrum of values, respond appropriately in light of both values and the best available scientific evidence, and find a path forward for the accomplishment of their legal and professional mandates. O'Brien (2003) notes several reasons why decision makers and managers have found it difficult to include public values in environmental decisions, including the difficulty of obtaining information about public values, the challenge of systematically incorporating diverse values into decision making, and professional views about the knowledge, practicality, and stability of public attitudes.

This litany of reasons points out the challenges that must be overcome if the demand for public involvement is to be met in a manner that both accomplishes the purposes for public involvement and enhances the quality of environmental decision making. As Gregory and Wellman (2001) point out, the challenge is to find a way to involve the public meaningfully at a detailed, action-specific level, while at the same time ensuring that the judgments made are informed by and recognize the complex, multidimensional nature of the initiatives under consideration.

To meet this challenge it is helpful to first consider values within an "appreciative system" as described by Vickers (1965). To present Vickers's appreciative system most succinctly, we follow the systems description developed by Checkland and Casar (1986) and presented graphically (see fig. 2.1) in simplified form in Checkland (2005).

This representation is essentially a constructivist view, which argues that people interpret and make judgments about their life experiences (1) in light of past experience (2) and their standards, norms and values (3). Based on these judgments (4), they take action (5) to influence their real-world experiences. Values are not static, but rather are influenced by our past socialization and relationships, our life experiences, and by public discourse (O'Brien 2003). While this model can most readily be conceptualized at the individual level, Vickers applied the same model to groups, organizations, and society as a whole.

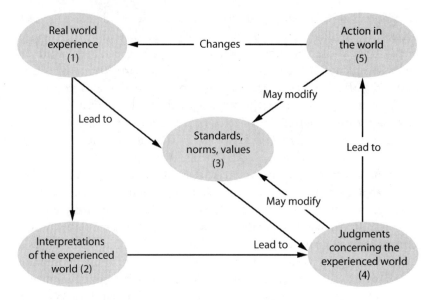

FIGURE 2.1. An appreciative system
Note: Based on the work of Vickers (1965); figure adapted from Checkland (2005)

We will have more to say about Vickers's model and its links to the personal construct psychology of George Kelly later, but for now the key point is that values play a vital role in framing problems. Values inform our judgments about what is problematic. Values have hierarchical relationships to other values and thus allow us to rank competing values within a given decision context. This ranking then guides our choices among alternative policy goals. What should be clear at this point is that, in the context of addressing wicked problems, both public discourse or dialogue and solid science are essential so that values of all parties can be illuminated and decision makers can learn from each other, which may modify values. With both the prevalence of uncertainty and the importance of values identified, it is also important to understand how the two relate and provide a foundation for understanding risk.

Risk

Risk is a compound measure of the probability and magnitude of an event or adverse effect (Dietz, Frey, and Rosa 2001). While risk can be measured in terms of probabilities and magnitudes, unlike uncertainty classifying something as *risky* is a social construction, a value judgment associated with an event of a known probability. Society determines which risks are important and should be attended to. Different individuals and societies perceive risks differently, and one problem is that society cannot but ignore the *we do not know that we do not know* type of risks; therefore, our perceptions of what is risky and whether we should do something about it vary across space and time. As mentioned earlier, different social science disciplines provide us with varied insights into how and why these different perceptions of both uncertainty and importance arise.

The Economic Perspective

Economic risk analysis does not concern itself with the social construction process of risk perceptions, but focuses on risk measurement, evaluation, forecasting, controlling, and decision making, using economic techniques and methods. Thus, the definition of risk is mainly technical, such as "risk is the variation in outcomes around an expectation" (Fone and Young 2000).

The economic approach operates as if the economic value (cost) of risk can be accurately measured. This assumption is this perspective's major problem since benefits or costs cannot be measured directly through monetary units, especially when considering social and ecological costs or benefits. Although the validity of such preference revelations is hotly debated, economists argue that they are more reliable than any other form of direct measurement (Fisher and Chen 1996). Economic decision making linked to risk has two steps: to identify the possible consequences of action or nonaction, and to choose the action based on comparing these consequences using some preestablished criteria. In reality, the knowledge of alternatives, consequences, and probability distributions is far from complete, so rational decision making can only be incomplete and fragmented (Simon 1960). Generally, with the economic approach decision makers concentrate on analyzing possible monetary loss, or on analyzing the comparison between costs and benefits, using methods like net present value, cost-benefit analysis, impact analysis, or the relatively recent value-at-risk analysis.

Applying economic risk analysis in public decision making is complicated and may be misleading. Some of this difficulty arises from the fact that for multigoal decision making, decentralized decision making, and decision making that involves mainly nonmarketable public supplies and services, the economic perspective requires consistent transformation of risk perceptions into comparable quantities, and the ability to pool such perceptions.

The Psychological Perspective

Experiments involving risky decisions have demonstrated that most individuals systematically under- or overestimate risks (Tversky and Kahneman 1982), which may reflect fundamental processes in the cognitive organization of risk perception. A number of commonsense strategies or cognitive rules of thumb apparently produce these biases. Many of these rules of thumb have evolved out of psychological experiments on how humans perceive risk and behave in risky contexts (Lichtenstein et al. 1978). These researchers have found that we perceive losses differently than gains and often we focus on the change rather than on the total amount. Hence, investors' sense of how well they are doing depends more on the returns than on the size of the assets.

Why are laypersons' judgments sometimes correct, but more often incorrect? What cognitive processes could produce such contrary results? Fischhoff and colleagues (1981) and Tversky and Kahneman (1982) propose that individuals employ a number of shortcuts when making decisions, and many of these heuristics have built-in and known biases (Combs and Slovic 1979; Sandman et al. 1987). Because of these biases it is difficult to identify underlying preferences. Consider, for instance, how our preferences change depending upon how the issue is framed.

Tversky and Kahneman (1982) noted that, when offered a choice between a sure gain of $240, a 25 percent chance to gain $1000, and a 75 percent chance of getting nothing, more than four-fifths of people chose the sure gain. However, when offered the choice between a sure loss of $750, a 75 percent chance of losing $1000, and a 25 percent chance to lose nothing, almost three-quarters chose to gamble. This experiment suggests that when an alternative is framed as a gain, people are attracted toward the sure gain. However, when the problem is posed in the form of a loss, people seem to prefer the gamble to the certain, potentially smaller, loss.

Discrepancies between public and expert perceptions of risk are seen in information about risk, and in laypersons' tendencies to stress qualitative features of risk ignored by experts (Slovic 1987). In fact, discrepancies in lay and expert judgments of risk are based on different definitions of risk (Dietz, Frey, and Rosa 2001). If the public and the experts have different views of risk, of what value is public opinion to risk evaluation? And what is the proper role of the public in risk management?

Three positions are described in the literature. The first position argues that the public should be excluded from risk assessment and decision making (Starr 1969; Cohen 1987; Breyer 1993). The second position proposes that laypersons' perceptions of risk should be brought in line with the experts' (Covello, Sandman, and Slovic 1989). The third position recognizes that nearly all risk assessments and risk management strategies are laden with uncertainty, that experts as well as the public are subject to cognitive biases, and that an emphasis solely on technical information has political implications for the relative power of different interest groups. For these reasons, laypersons should play a more central role in the process of assessing, evaluating, and managing technological risks (Perrow 1984; Dietz 1987; Freudenburg 1988; National Research Council 1989; Fischhoff 1990; Brown 1992; Rosa and Clark 1999). Unfortunately, public participation does not always ensure agreement on the perception and relative importance of the risks.

The Sociological Perspective

The sociological perspective seeks to understand social influences on risk perception and behavior, the importance of organizational contexts and institutional responses to risk, and the role of risk in large-scale social change. Four distinct research directions, representing increasing levels of theoretical aggregation from micro to meso to macro, are described according to this perspective.

The first direction builds upon psychometric research, but its goal is to reconceptualize the psychometric findings through a sociological lens (Dietz, Frey, and Rosa 2001). The second sociological research direction fundamentally reconceptualizes the psychometric model by proposing a model that examines the problem of risk perception by taking into account the social context in which human perceptions are formed (Rosa, Mazur, and Dietz 1987). This perspective notes that people often take action or form attitudes prior to developing meaningful perceptions. The psychometric model ignores this fact, but it is a central feature of this sociological model.

The organizational and institutional approach emphasizes the system characteristics, the context of complex issues (environment, technology), and the policies that develop for their use. According to this direction, risks can only be understood by analyzing the way parts of risky systems fit together (Freudenburg 1988)—a sociological focus on the organizational and institutional contexts of decision making.

The fourth line of thinking deals with worldwide social change: the transformation from modernity to its successor, some form of postmodernity. Risk is the central driving force of this transformation (Giddens 1990).

Modernity results in globalization. Global interdependence grounded in shared risks magnifies the importance of trust. Giddens's argument is similar to Beck's (1992) "reflexive modernization," and distributing "bad" dangers, which form the basis of "the risk society." This social change results in a decline in the importance of structures, and the individualization of social agents who, forced to make decisions, reflect on the social institutions responsible for those decisions. Like Giddens, Beck defines risk globalization and emphasizes the role of trust in dealing with these risks. In addition, Beck underscores the role of science in issues of risk, a facet emphasized by other scholars as well (Dietz, Stern, and Rycroft 1989; Brown 1992; Burns and Dietz 1992). On one hand, science is partly responsible for the growth of risks and hazards; on the other hand, science is the principal social institution entrusted with knowledge claims about risk. Since science is no longer privileged, risk societies make knowledge claims about the increased risks defining societies through Beck's "reflexive modernity"—that is, a negotiation of knowledge claims between science, political interests, and laypersons. This reduced status for science, and the relative increase in the ability of different community groups to form coalitions that challenge administrative action, is a relatively new phenomenon. We have yet to develop the institutions and mechanisms that provide the stability and trust necessary for developing and implementing approaches to addressing complex long-term problems.

The Anthropological Perspective

The anthropological approach to risk analysis incorporates cultural theory. In contrast to the psychological perspective, but in common with the sociological perspective, the anthropological approach emphasizes shared beliefs and values, or the cultural construction of cognition, in determining not only the problem definition, but also means/end perception and expost evaluation. Risk can be linked to culture in multiple ways (Marsella 1998).

Douglas and Wildavsky (1982) argue that selecting risks for societal attention is purely a social process with little or no linkage to objective risk or physical reality. Social disagreement over defining risks is the fundamental source of uncertainty, and the multiple and dynamic characteristics of culture make it impossible to reach a full compromise in terms of cultural cognition in almost every public policy.

Swedlow (2002) reframed the typology proposed by Douglas and Wildavsky (1982) to identify four patterns of cultural biases and how they relate to basic risk understanding and risk management strategies. The *individualistic* pattern values liberty, autonomy, and equilibrium through dispersed behavior. Thus, risks stem mainly from bounded individual rationality and from undue intervention of central control. The *fatalistic* pattern relies on random luck and demands merely survival: risks stem from uncontrollable contingencies. The *hierarchical* pattern values order and collective acts under a central plan: risks result from absence of concentrated coordination. Finally, the *egalitarian* pattern values equality, not just within the human community, but also between the social and natural communities. These patterns may well shape how individuals interpret potential problems and their policy implications.

While various groups are segmented by their opinions on risk perception of the same public issues, they generally come to a consensus with the cultural risk of decision making. This type of risk is called a risk of consensus, or the risk of legitimacy. Yet, such cultural consensus cannot be easily achieved by political compromise, because culture may not constitute a conscious variable in decision-making processes. The cultural environment is dynamic, and its influence on risk perceptions—including identifying and defining problems, assessing their seriousness and the maximum risk level society can tolerate, considering the proper ways to mitigate risk, and evaluating costs and benefits—also will be dynamic and mixed.

Because there is no absolute criterion for judging and ranking cultural stances, risk aversion implies making open-ended decisions that try to incorporate every value orientation and to allow for the possibility of expanded application. For complex situations with multiple cultural backgrounds involved, such pragmatic open-ended decision making takes longer and brings about excessive cost.

In a democratic society, slow decision making coupled with the potential for changing political support creates uncertain and unstable environments in which administrators need to make decisions that have long-term implications for implementation and consequences. The risks associated with such administrative uncertainty are often referred to as *administrative* risk.

The Spatial Perspective

In the field of geography, researchers traditionally explore human responses to natural disasters and human activities in a spatial context (Dietz, Frey, and Rosa 2001). The *social amplification* framework borrowed from communication theory attempts to link risk perception and behavior (Machlis and Rosa 1990). Risk events are signals that are amplified before reaching the ultimate receiver: the public. The amplification may be either heightening or attenuating the risk. Since amplifying risk signals is due to cognitive heuristics, the framework attempts to bring together psychometric findings on risk perception with the institutional context of risk communication in order to better predict responses to risk.

The geographic perspective emphasizes issues of location and place, thereby highlighting yet another factor that must be taken into consideration when making decisions. Local communities often attach special meanings to specific locations (Putnam, Leonardi, and Nanetti 1993). When decision makers lack knowledge about local sensibilities, the uninformed decisions can cause serious problems when implemented.

Integrating the Multiple Perspectives

These different social science perspectives on risk are, by definition, only illustrative of risk's complexity as a theoretical notion, and its operationalization in the different disciplines. Our summaries do not, by any means, exhaust the multiple perspectives and their obvious and subtle differences. The lesson to be drawn from these brief sketches is that even in the simplest decisions there are a number of theoretical issues to take into account when making decisions under uncertainty.

For public decision making, deciding parties would be prudent to first analyze the direct costs and benefits associated with risks, to the extent they can be estimated, and then to identify the possible social consequences together with their implied value(s). Public deliberation about these consequences and their valuation could potentially lead to improved decisions.

The costs and benefits associated with risks depend upon how the risks are perceived. How people perceive risk depends on (Tversky and Kahneman 1982) what they value, how the risk is framed, and their level of trust in the responsible organization or institution. It is well known, for example, that there is an inverse relationship between perceived risk and perceived benefit, and this relationship is linked to an individual's general affective evaluation of a hazard. If an activity is *liked*, people tend to judge its benefits as high and its risks as low. If the activity is *disliked*, the judgments are the opposite—benefits tend to be perceived as low while risks are perceived as high (Slovic 2000).

Further, and perhaps even more important, every way of presenting risk information is a *frame* that can shape participants' judgments in a risk decision. If the issue is framed in a positive light, people are more likely to dwell on the decision's positive aspects, and vice versa. One often-cited example is the observation that summarizing medical risks in terms of mortality rates yields very different perceptions than when the same information is presented in terms of survival rates. If a given treatment is described as having a mortality rate of 10 percent, for example, it is perceived very differently than if the same treatment is said to have a survival rate of 90 percent (Tversky and Kahneman 1982). Evidence also shows that experts are not immune to these framing effects. The effect is as strong when subjects are physicians as when they are lay people (Watzlawick, Weakland, and Fisch 1974). As a report of the National Research Council (1996, 57) concludes, "Numerous research studies have demonstrated that different but logically equivalent ways of summarizing the same risk information can lead to different understandings and different preferences for decisions."

Note that this is not an issue that can be resolved with better science, because there is no scientific way to determine that one risk summary is more accurate or less biased than another when both accurately reflect the same data. Consequently, the problem of generating a single unbiased risk information summary to meet participants' needs in a risk decision has no purely technical solution. The problem is further complicated by the fact that there are many sources of uncertainty that give rise to multiple risks. Natural disasters are risky for humans and for the natural habitats of various animals, birds, and other creatures. They also pose a risk to streams, trees, and other vegetation. Thus regardless of how we choose to value them, we have no consensus on the potential consequences of a natural disaster.

As with uncertainty, to resolve this type of dilemma one must focus on the employed decision processes. In this light, it is also important to note a corollary to the affective evaluation principle mentioned earlier: if participants trust the organization presenting the risk information, they are more likely to accept the characterization. That level of trust is a byproduct of the decision process. Experience in a variety of settings suggests that such trust is easily damaged and difficult to restore (Douglas 1985). As we will elaborate later, because of uncertainty about outcomes and the difficulty in predicting the consequences of taking a particular course of action, it is important to implement a process that engenders trust. With such trust, it is easier to accept the actual outcomes as being fair even though they may not be what one had hoped for.

Rittel and Webber's (1973) ten propositions describe characteristics of a wicked problem; however, they do not provide a test for determining whether a problem is wicked. Are there necessary and sufficient conditions for determining whether a problem is wicked? Not all problems with multiple stakeholders and uncertain outcomes are wicked. In fact, the regional forester for the Sierra Nevada forests successfully managed the forests for many decades without being stymied into a stalemate brought about by conflicting interests and irreconcilable differences.

Addressing Wicked Problems

So, what makes a problem wicked? Can we articulate a set of preconditions, which when present, imply that a policy concern or issue will be intractable? The essential ingredients of a wicked problem are difficult to pin down. Although there are no sufficient conditions that serve as a test for identifying wicked problems, there are some conditions that generally accompany them. The conditions listed in table 2.2 are adapted from Rittel and Webber's propositions generally associated with wicked problems.

This chapter began with the question, what makes some decisions more difficult than others? However, in dealing with wicked problems, the question is not one of deciding on which criteria to employ and how to use those criteria to select among a set of possible options. The issue is more complex. It requires that the decision makers first have a common understanding of the situation they are attempting to address. That entails establishing boundaries to ascertain what lies within and what is outside the problematical situation, determining who the stakeholders are and clarifying their values.

According to Vickers (1965, 40), decision makers need to develop an "appreciation" of the problem by which they reach "judgments of fact about the 'state of the system' both internally and in its external relations" and "judgments about the significance of these facts to the appreciator or to the body for whom the appreciation is made." For Wildavsky (1979) this appreciation, which eventually leads not to a solution, but some resolution or settlement, requires "a smart mix of cogitation and interaction" (Grin and Hoppe 2000, 180). Conklin and Weil (1997) suggest that a characteristic of wicked problem appreciation is that "you don't understand the problem until you have developed a solution." Thus, despite increasing experience with such problems, the definition of wicked problems remains largely unchanged after almost forty years.

Although the definition has not changed, we now understand the need to think of wicked problems in a holistic fashion so as to be able to take into consideration the interacting and interdependent parts of the

TABLE 2.2. Conditions associated with wicked problems (Rittel and Webber 1973)

	Condition	Explanation
1	Lack of an unique problem statement	Multiple stakeholders have multiple perspectives on the problem resulting in lack of clarity regarding the nature of the problem.
2	Conflicting objectives	Because success is generally determined in terms of objectives, any ambiguity in purpose leads to lack of clarity about successful outcomes.
3	Conflicting values	Values determine the criteria by which success is to be judged, so any ambiguity in these criteria leads to lack of clarity about successful outcomes.
4	Dynamic context	Static solutions do not work in a dynamic context where problems are changing or evolving.
5	Scientific complexity and uncertainty	Uncertain or incomplete knowledge leads to an inadequate basis for informing decisions.
6	Political complexity and uncertainty	Ambiguity about political coalitions and power results in lack of clarity about which values should dominate and prevail.
7	Administrative complexity and uncertainty	Ambiguity about budgets and lack of procedural continuity results in inadequate implementation.
8	Multiple tactics to address problems	Lack of clarity about objectives and criteria for judging success result in lack of clarity about how best to proceed.
9	Multiple stakeholders with the power to assert their values	Multiple stakeholders with multiple value sets and power structures lead to conflicting definitions of success.

system that underlie the problematical situation (Checkland and Poulter 2006). However, that is easier said than done, for it involves having a deep understanding of the context and an appreciation for the multiple perspectives, interests, attitudes, and values that the multiple stakeholders bring to the decisions. To provide a flavor of the complexities involved, we will describe four different contexts in which decisions have been stalled because of the difficulties involved in dealing with the wickedness of the underlying policy situation.

Chapter 3

Four Wicked Cases

In this chapter, we discuss in more detail the four major environmental controversies introduced in chapter 1. These four examples—restoration of the Everglades, management of Ngorongoro Conservation Area in Tanzania, carbon trading in the European Union, and management of the Sierra Nevada national forests—illustrate the breadth of environmental challenges that fall into the category of wicked problems. The cases involve a range of socioeconomic contexts in both developed and developing countries and center on a variety of natural resources. Forest planning in the Sierra Nevada will be our primary case study throughout the book, but we will draw on the other three cases frequently to illustrate our points.

While these four cases share the general characteristics of wicked problems, they differ in the details. The current effort to restore the Everglades attempts to balance conflicting demands linked to agriculture, residential and commercial development, water supply, flood control, and protection of a major national park. In Tanzania's Ngorongoro Conservation Area the dilemma is to integrate wildlife conservation, international tourism, and traditional Maasai livelihoods. The central challenge for the European Union's cap-and-trade program for greenhouse gases is to achieve strict emissions targets while maintaining the support of industry and the public. Finally, in the Sierra Nevada the US Forest Service struggles to develop management plans for the national forests that can win broad support despite the active engagement of powerful stakeholder groups with a wide range of competing values and priorities, including timber harvesting, recreation, and biodiversity protection.

While the details differ, these cases share important traits that reflect the dilemmas inherent in wicked problems. Planning horizons are long term and large scale. Underlying ecological and social conditions shift over time. Multiple and compounding uncertainties make it impossible to predict the outcomes of policy interventions with confidence. And competing private interests and divergent public values preclude consensus and destabilize coalitions. In all four cases, the goal of developing and implementing politically acceptable, technically feasible, and ecologically and economically effective policies seems unattainable. Yet inaction is also likely to have significant negative consequences. Our aim in this book is to explore possible responses to the challenging dilemmas embodied in these and other wicked environmental problems.

The Everglades

Florida, south of Lake Okeechobee, is low lying and flat. A narrow coastal ridge extending from five to twenty miles inland from the Atlantic shoreline at the peninsula's southern end rises to elevations of twenty to twenty-five feet and provides a geologic foundation for the region's major metropolitan development. A more modest ridge also rises along part of the western side of the lower peninsula. But much of the region's interior is no more than ten feet above sea level. The terrain slopes gently down across the hundred miles from the southern lip of Lake Okeechobee to Florida Bay. The area typically receives about sixty inches of precipitation a year, most falling during the summer rainy season. The saturated underlying limestone of the drainage basin retains water through seasonal cycles of precipitation. Given its topography and climate, the region in its natural state encompassed a broad aggregation of wetlands through which a wide, shallow sheet of water flowed slowly southward (Walker and Solecki 2001).

The hydrological system begins about one hundred miles north of Lake Okeechobee, just south of what is now Orlando, where the outflow from a series of lakes forms the Kissimmee River. Before modern development, the river followed the tortuously winding path common to flatland rivers. During rainy periods, the Kissimmee overspread its banks, and Lake Okeechobee spilled water over its southern lip into the vast wetlands of

South Florida. The boggy, impassable six thousand square miles of sawgrass marshes, cypress swamps, and coastal mangrove forests of the lower peninsula resisted major human development until modern times.

After American independence from Britain, Spain retained possession of Florida, but ultimately ceded the region to the United States in 1821. Conflicts between the American government and Florida's Indians continued for the next forty years, in a struggle known as the Seminole Wars. By the mid-nineteenth century, the government had driven most Indians out of northern Florida, shipping many to reservations in the American West. But several bands of Seminoles and other tribes continued to find refuge in the impenetrable Everglades (Walker and Solecki 2001).

The first serious effort to drain and develop South Florida began after the Civil War (Grunwald 2006). In 1881, Hamilton Disston, a real estate developer, purchased four million acres of land in the upper reaches of the drainage basin and signed contracts to drain eight million acres more. The state was eager to make the deal to help retire its postwar debts. Following Disston's investment, Henry Flagler, a former partner of oilman John D. Rockefeller, began to buy land and build railroad lines and tourist hotels along the Atlantic coastal ridge.

After a decade and a half of dredging and canal building, Disston was able to drain only parts of the Kissimmee valley north of Lake Okeechobee, a small fraction of the enormous area he originally hoped to dry out. Nevertheless, Disston's and Flagler's involvement triggered rapid regional development. Farmers began to grow sugarcane in the drained Kissimmee valley and plant citrus orchards along the western side of the railroad lines running down the Atlantic coast. Developers marketed the drained swampland widely and aggressively. By the turn of the twentieth century, immigration around the periphery of the Everglades was accelerating, land values were rising, and new cities were growing rapidly (Grunwald 2006).

In the Progressive Era of the late nineteenth century and early twentieth century, conflicts over the Everglades emerged that foreshadowed the struggles that continue today. Even before major drainage efforts took hold, these early conflicts centered on questions of what the land should be used for, who should pay for reclamation, and who should benefit. Progressive reformers advocated sustainable use of land, wildlife, water, and other resources, as opposed to wholesale exploitation on one hand or permanent wilderness protection on the other. But to Progressives *progress* ultimately meant productive use, and productive use of land in South Florida meant drainage (Grunwald 2006).

Napoleon Broward, a Progressive politician elected governor in 1904, set in motion ambitious plans to extend Disston's earlier efforts. With the goal of controlling the level of Lake Okeechobee and cutting off the main source of water for the Everglades, the plan included the construction of a dike around the southern end of the lake and the dredging of a series of canals to drain water east and southeast through the Atlantic coastal ridge to the sea. Between 1912 and 1926, Broward's successors oversaw the completion of the dike and six drainage canals.

But these projects failed to achieve the goals of managing water flows and opening all of South Florida to development. In the late 1920s two major hurricanes caused severe flooding and destruction and exposed the limitations of the work that had been done. In response, the US Army Corps of Engineers began to take on greater responsibility for environmental management. In the 1930s, the Corps enlarged and strengthened the dike at the southern edge of Lake Okeechobee and deepened a canal that Disston had built along the channel of the Caloosahatchee River to take water from the lake out to the Gulf of Mexico.

Commensurate with the area's growing population and economy, the scale of engineering interventions to manage water flow grew substantially in the 1940s. Congress in 1948 authorized the Army Corps of Engineers, under the sweeping Central and Southern Florida Project, to control flooding, ensure water supplies, and otherwise promote development (Kiker, Milon, and Hodges 2001). The plan established agricultural development zones in the Kissimmee valley and in a large area abutting Lake Okeechobee to the south. This territory, known as the Everglades Agricultural Area (EAA), encompassed about 25 percent of the original Everglades region. Most of the Kissimmee valley became grazing land for cattle, while sugarcane plantations came to dominate in the EAA. The Central and Southern Florida Project allocated another 40 percent of the original Everglades, lying to the south of the EAA, as water conservation areas, where dikes, canals, pumping stations, and reservoirs would provide regional flood control and water management. The project also envisioned space for suburban and agricultural development in the eastern Everglades on the interior side of the growing cities of the Atlantic coastal ridge (Grunwald 2006).

Separately, in 1934, Congress approved setting aside the southernmost 25 percent of the original Everglades as a national park. Everglades National Park was finally dedicated in 1947 after the state legislature provided funds to acquire the necessary land. Big Cypress Swamp, a large tract of wetlands on the western side of the peninsula bordering Everglades National Park, remained in private hands until 1971 when the federal government purchased the property and created Big Cypress National Preserve.

From the 1950s to the 1980s, the Army Corps of Engineers undertook a massive effort to implement the Central and Southern Florida Project. The Corps dredged a deep channel to straighten the Kissimmee River. The agency also built levees completely around Lake Okeechobee, around the newly established Everglades Agricultural Area to the south of the lake, and down the western side of the Atlantic coastal ridge to open that area for suburban and agricultural development. The Corps also constructed major water control infrastructure in the water conservation areas, establishing reservoirs, expanding the capacity of existing canals, and adding pumps and controls throughout the system to manage water flow. The construction of two highways across the Everglades—US 41 (the Tamiami Trail), completed in 1928, and Interstate 75 (Alligator Alley), completed in 1969—also significantly affected the region's hydrology and ecology.

Repercussions of Development

In retrospect, the enormous hundred-year effort from the 1880s to the 1980s to dredge, drain, reclaim, develop, and manage South Florida led to decidedly mixed results (Grunwald 2006). Large areas were opened for productive agriculture, and the region's demographic and economic boom expanded dramatically. Florida now has the fourth largest population and fourth largest economy among US states. Land reclaimed from the South Florida drainage basin accounts for about half the US sugarcane harvest (Baucum and Rice 2009) and two-thirds of domestic citrus production (USDA Economic Research Service 2007). Almost six million people now live along the eastern edge of the Everglades in the Miami metropolitan area (Kranzer 2003).

But serious problems also arose. In the early days, many land sales were outright swindles and scams, with the proffered sites still under water or inaccessible. Some of the new lands that could usefully be cultivated quickly lost their fertility. And once drainage programs took hold, water tables dropped; dried out areas suffered recurring floods, fires, and dust storms; soil subsided; canals silted up; salt water infiltrated the groundwater; and the growing cities of the Atlantic coast began to experience seasonal water shortages (Grunwald 2006).

By the late 1980s, following massive investments in the Central and Southern Florida Project, many of the old problems had become more severe and new problems had emerged. Moreover scientists, policy makers, and the general public had become more aware of the scale and interconnectedness of the challenges. Demographic and economic expansion in South Florida over the previous century had brought many benefits to the people who lived and worked there, but political leaders and experts at the state and federal levels had underestimated and ineffectively managed the environmental damage associated with the development. The damage had become extensive enough to threaten the regional economy.

By this time, half the original Everglades was gone, and what remained was degrading rapidly, both as an ecological reserve and a source of fresh water, recreation, and other environmental benefits (National Research Council 2006). Threats to the Everglades followed directly from the engineering interventions undertaken during the previous decades. The extensive channelization and diking of the upper drainage basin isolated what was left of the Everglades from its natural sources of water recharge. Perhaps more damaging, the water it did receive was polluted by agricultural runoff.

The natural Everglades system is phosphorus limited, so farmers used artificial fertilizers to support the successful production of crops, particularly sugarcane (Walker and Solecki 2001). Excess nutrients from fertilizer applications washed down into the Everglades from the agricultural area and caused widespread damage, changing soil chemistry, killing off invertebrates and small fish at the bottom of the food chain, replacing native sawgrass with cattails and other invasive plant species, and destroying habitat for fish and waterbirds. In the 1980s, biologists documented these changes spreading out and down from the agricultural area as far as the Everglades National Park at the southern end of the peninsula.

As indicators of environmental stress in the Everglades, the huge flocks of waterfowl that previously inhabited the region were gone, and the Florida panther was nearly extinct (Walker and Solecki 2001). But the adverse effects extended beyond the biodiversity of the swamps and marshes. As canals and dikes concentrated and accelerated the flow of water through the system, pulses of tainted water created dead zones in Florida Bay and along the eastern coastline (Bhat and Stamatiades 2003).

The reengineered hydrology of South Florida also contributed to water shortages and declines in water quality (Walker and Solecki 2001). Pollutants contaminated the water in the channels thereby increasing health risks, and water that flowed rapidly out to sea during rainy seasons was lost as a resource for residential, commercial, and agricultural uses (Sheikh and Carter 2008). Moreover, despite the massive water management infrastructure, the region continued to experience floods and droughts.

Powerful constituencies competed to shape the responses to these problems. Strongly active stakeholders included sugar plantation owners, environmentalists, Indian groups, and government officials. The issue also drew national attention. Florida had come to play an important role in national electoral politics, and the Everglades National Park was a symbol of conservation, or conservation failures. Presidents and presidential candidates, along with senators from Florida and other states, worked to turn the political controversy over environmental management in South Florida to their advantage (Grunwald 2006).

Recent Mitigation Efforts

Reflecting global interest in the Everglades, the United Nations Educational, Scientific, and Cultural Organization (UNESCO) listed the national park as a World Heritage Site in 1979. In 1993, the UN's World Heritage Committee pointedly added the Everglades to its list of threatened sites, causing the United States considerable embarrassment. In 2007, acting on an American request, UNESCO removed the Everglades from its List of World Heritage in Danger despite the limited effectiveness of measures to improve the park's prospects (Pittman 2007).

In the 1990s, battles over the Everglades escalated, but at the same time the intensity of the struggles moved regional conservation and water management to the top of the policy agenda and created opportunities for compromise. In 1992 and 1994, the US Congress and the Florida legislature enacted several initiatives to address the interrelated problems of water supply and ecological restoration in South Florida. These projects included dechannelizing the Kissimmee River and installing filter marshes around the Everglades Agricultural Area to clean agricultural runoff before pollutants reached Lake Okeechobee and the Everglades. Federal and state agencies shared the costs of land acquisition required to begin these projects (National Research Council 2006).

The 1992 legislation also authorized the Army Corps of Engineers to conduct a detailed review of the Central and Southern Florida Project. This review culminated in a report that was radical in its assumptions and recommendations. It led to development of the Comprehensive Everglades Restoration Plan (CERP), an exceptionally ambitious and expensive mitigation program. The CERP, originally priced at about eight billion dollars over forty years, envisioned a massive new water management effort that would provide sufficient water to serve both the rapidly growing South Florida economy and the Everglades National Park. While the Corps would continue to engineer the hydrological system of the vast drainage basin, it would also try to restore, to the degree possible, the natural, broad, shallow, slow-moving sheet of clean water flowing south from Lake Okeechobee to Florida Bay that had existed before implementation of the Central and Southern Florida Project. In November 2000, Congress passed the bill authorizing the CERP, and President Clinton signed the bill into law a month later. The plan called for the state and federal governments to share the costs, and the Florida legislature passed laws authorizing the state's participation.

This brief summary of the emergence of the Comprehensive Everglades Restoration Plan grossly understates the struggles of its birth. The decision-making environment had all the characteristics of a wicked problem. Many powerful stakeholder groups with contradictory priorities and divergent values battled to shape policy (Grunwald 2006). Beyond their mutual disagreements, many of these constituencies had substantial intragroup conflicts. Among government agencies, for example, the Corps struggled internally regarding both its past actions and its future responsibilities. The National Park Service was sharply critical of the Corps' initial version of the CERP. And state and federal authorities argued over the role of the US Interior Department.

Environmentalists were also divided among themselves (Grunwald 2002; Clarke and Dalrymple 2003). Some groups, led by the Audubon Society, took the position that the price of inaction on the Everglades was so high that policy compromises to get the CERP enacted were worth the price. Others, including the Environmental Defense Fund and the Natural Resources Defense Council, held out for stronger protections for the national park and endangered species, even at the risk of scuttling the plan altogether.

Moreover the political environment of 1999 and 2000, during the run-up to the congressional vote on the CERP, was particularly tangled and daunting at both the state and federal level. In January 1999, the Florida governorship passed from Lawton Chiles, a Democrat who had taken the lead in reforming regional water management policies, to Jeb Bush, a Republican former land developer. In October of that year, the unexpected death of Senator John Chafee of Rhode Island, a proenvironmental moderate, transferred the chairmanship of the critically important US Senate's Environment and Public Works Committee to Robert Smith of New Hampshire, an antiregulatory conservative. Moreover, the final vote on the CERP in Congress occurred after Election Day in 2000, in the midst of the furor over the Florida presidential election recount.

Even to get to the vote on the Comprehensive Everglades Restoration

Plan, participants had to juggle a combustible set of subproblems nested like Russian matryoshka dolls within the overall dilemma of South Florida environmental management. In 1996, for example, Dade County approved plans to convert part of Homestead Air Force Base to a commercial airport. This ignited a fight between advocates for economic development and environmental protection that reprised a battle in the late 1960s over a proposed airport in the Big Cypress Swamp. In both cases, the president, Nixon in 1970 and Clinton in 2001, ultimately decided against the proposed airports. But the political struggle in the late 1990s and into the 2000s over the Homestead airport continually threatened to break up coalitions favoring the CERP.

A second example centers on an area known as the Eight-and-a-Half-Square-Mile Area. This was a rural development, home to about 350 mostly Cuban American families, carved out of the Everglades on the west side of the dike protecting developed areas on the Atlantic coastal ridge. Given the relatively small size of the area, the high cost of protecting it from flooding, and its importance to the Everglades restoration effort, Governor Chiles had supported a plan to buy out the residents. In 1999, Governor Jeb Bush overruled this option, putting restoration efforts near the area on hold.

The CERP also faced deep scientific uncertainty. The hydrological and ecological systems of South Florida are highly complex, certainly beyond the ability of scientists and engineers to model and manage with confidence (Gunderson 2001; Sklar et al. 2001). Moreover, unexpected future social and economic changes are sure to affect outcomes in unpredictable ways. Finally, rising sea levels linked to climate change may radically alter the region's geography, overwhelming restoration efforts (Wanless, Parkinson, and Tedesco 1994). Despite these daunting challenges, a resilient coalition of unlikely bedfellows from the state and across the nation survived the political turmoil of the late 1990s and early 2000s long enough to enact this ambitious and expensive plan to address South Florida's environmental problems.

Current Situation

Yet, in the implementation, what appeared to be remarkable progress toward taming an intractable wicked problem faltered (U.S. Government Accountability Office 2007). The state under succeeding administrations has continued to purchase land to implement the Comprehensive Everglades Restoration Plan, install filter marshes to mitigate phosphorus pollution, and begin construction of water supply reservoirs. The most dramatic sign of Florida's continuing commitment to the CERP occurred in mid-2008 with the announcement of a tentative $1.8 billion deal to buy out the U.S. Sugar Corporation's operations and land holdings in the Everglades Agricultural Area (Cave 2008). If this complex deal is ultimately completed, it would both eliminate a major source of pollution in the Everglades and substantially increase the land available for restoration. Since 2000, however, Congress has repeatedly failed to appropriate the funds necessary to support essential federal action. Wars in Iraq and Afghanistan, a major nationwide financial crisis, and growing public-sector budget deficits have changed national priorities, and key supporters in the US Senate, including Mack and Graham from Florida, have retired.

Without full federal engagement, the state is severely handicapped in its efforts to complete the CERP. Moreover, critics continue to accuse the Army Corps of Engineers of inflexibility and a failure to institutionalize environmental stewardship (Goodnough 2007). And Florida itself has also failed to fulfill all its promises. State-sanctioned residential and commercial development in the Everglades moves forward, even in areas marked for restoration under the CERP. While farmers fall behind pollution reduction benchmarks, the state legislature postpones enforcement deadlines (U.S. Government Accountability Office 2007). Infrastructure projects under way are behind schedule, over budget, and focused primarily on economic development, while projects with a primarily environmental purpose are typically put off.

In sum, the inherent wickedness of the problem of the Everglades restoration remains untamed. Efforts to restore the Everglades have foundered at least in part on arguments over the relative priority of people and nature. Framing the debate in these terms helped the state justify its recent focus on projects intended to supply water for residential, agricultural, commercial, and industrial needs rather than on projects designed to maintain the viability of the national park and other nature reserves.

A long-term view suggests that a bright-line distinction between people and nature in South Florida is a false dichotomy. This view was most widely understood and articulated during the period of compromise leading to enactment of the Comprehensive Everglades Restoration Plan in 2000. The deterioration of the Everglades appeared to have reached a level of severity that generated adverse impacts on people as well as ecosystems and biodiversity. But later when policy makers had to allocate limited financial resources over the short term, they emphasized social and economic priorities over conservation. The majorities living in the area who favored people over nature had the political power to drive this choice.

Ngorongoro Conservation Area

The perceived dichotomy between people and nature also plays a central role in our second example of a wicked problem: management of Tanzania's Ngorongoro Conservation Area (NCA). In this case, however, the local people, primarily Maasai, get short shrift. Policies have emphasized protection of wildlife and habitat—and associated tourism income—over local community development (Homewood and Rodgers 2004).

The NCA, situated about a hundred miles west of Mount Kilimanjaro and fifty miles south of the Kenyan border, encompasses 3,300 square miles of Tanzania's dry northern highlands. The larger Serengeti National Park borders the NCA to the west, and the Great Rift Valley escarpment forms its eastern edge (United Nations Environment Programme 2008). The region supports large populations of wildebeest, zebras, giraffes, lions, leopards, hyenas, warthogs, buffalo, and many other wildlife species. Each year more than a million wildebeest, accompanied by hundreds of thousands of zebras and gazelles, migrate in seasonal cycles across the protected areas of northern Tanzania and southern Kenya following the rains to seek water and pasture.

The NCA is named for its most striking feature: Ngorongoro Crater, an enormous caldera with crater walls over a thousand feet high encircling a hundred-square-mile area of grassy savanna and open woodland in the collapsed crater. The floor of the crater, at about seven thousand feet above sea level, is home to high concentrations of ungulates, along with the lions, leopards, cheetahs, hyenas, and other predators they support.

Before European colonization, northern Tanzania was the territory of Maasai pastoralists, who in their customs valued seminomadic cattle herding, independence, and a warrior mentality, and devalued both sedentary agriculture and wild game hunting (McCabe 2003). The Maasai migrated into what is now Kenya and Tanzania from the north beginning in the sixteenth century, displacing indigenous groups. By the mid-nineteenth century, they had expanded their area of control to include all of what is now the Serengeti-Ngorongoro ecosystem.

At this time, Omani Arabs dominated the Indian Ocean coastal areas, sending slaves, ivory, and spices from East Africa to the Middle East and beyond. In the second half of the nineteenth century, Britain and Germany began to contest Arab power in the region, and British and German missionaries and explorers began to push into the interior. Outsiders, however, whether Arab or European, generally avoided the powerful Maasai.

Repercussions of Development

In the 1880s and 1890s, various catastrophes weakened the Maasai. A multiyear drought and recurrent locust infestations reduced the critical resources of water and pasture, outbreaks of cholera and smallpox decimated the Maasai themselves, and pleuroneumonia and rinderpest destroyed much of their livestock (McCabe 2003; Goldstein 2005). During the same period, Germany and Great Britain became more assertive in their efforts to control the East African interior. By 1890, the two European powers had negotiated spheres of influence, with Germany dominating most of Tanganyika, the mainland territory of what is now Tanzania, and the British establishing authority in Kenya and Zanzibar. In the first decade of the nineteenth century, the British evicted the weakened Maasai from the majority of the territory on which they had herded their cattle in Kenya to make way for white ranchers and planters and to establish game reserves.

After Germany's defeat in World War I, the British gained control of Tanganyika as well, as a protectorate under the auspices of the League of Nations. Here the British authorities also expropriated land from indigenous Africans, including the Maasai, taking territory both for settler agriculture and for safari hunting areas. The catastrophic decline of Maasai populations from disease outbreaks around the turn of the twentieth century left much of their lands underoccupied and thus more easily co-opted by British authorities (Goldstein 2005).

The colonial regime first established a game reserve in the core of the Serengeti in the late 1920s and formally declared the zone a national park in 1940. Over the next fifteen years, while permitting the Maasai to remain, the government promulgated a series of ordinances to restrict their activities. The government also extended the park's boundaries, adding substantial territory in the Serengeti savanna and incorporating the Ngorongoro highlands.

During this period, the Maasai and park managers faced off in continuous running conflicts over access to land and pasture. Finally in 1959 the colonial government addressed the problems as it understood them, separating the Ngorongoro Conservation Area from the Serengeti National Park. The new policy banned the Maasai from the Serengeti but allowed them access to the NCA for what were defined as their traditional activities of livestock grazing and small-scale subsistence cultivation. The new rules did not permit permanent settlements or wider cultivation. The policy established the NCA as a mixed-use zone integrating wildlife tourism and

Maasai pastoralism and created a new entity, the Ngorongoro Conservation Area Authority (NCAA), to manage the NCA.

Scholars have noted that several important assumptions underlie the game management and conservation policies implemented in Tanganyika and elsewhere in eastern and southern Africa (Neumann 1998; Goldman 2006). First, colonial authorities brought with them the concept of exclusive aristocratic access to game practiced in Europe. They also viewed humans as separate from nature, and internalized a myth of wild Africa largely untouched by human agency. Given these cultural norms, colonial authorities saw the presence of native Africans as threatening to natural landscapes and detracting from wilderness vistas and safari hunting experiences.

Second, the Europeans believed that native pastoralists had limited claim to the land because they had not improved it or managed it and were careless stewards of the environment. Colonial authorities assumed that indigenous livelihoods were incompatible with untrammeled nature and with effective wildlife management. These beliefs and assumptions justified removing natives from large nature reserves, even if local people had lived in the areas and used the natural resources in them sustainably for centuries. In this worldview, national parks such as Serengeti were necessary to preserve nature in a way that was scientifically and aesthetically appropriate.

Displaced locals, on the other hand, typically viewed this type of exclusionary conservation as dispossession, unfair dealing, and a de facto signal that wild animals had precedence over local people (Goldman 2006; Poole 2006). The colonial government of Tanganyika in the late 1950s saw creation of the mixed-use Ngorongoro Conservation Area as an enlightened way to manage the conflicts triggered by the establishment of parks that aimed to protect nature from locals. The Maasai, however, were unsatisfied, seeing the new policy as a bad bargain, imposed unilaterally, that constrained their independence, shrunk their territory, and restricted them to less fertile dryland areas.

External actors also played an important role in colonial Tanganyika's environmental decision making (Goldstein 2005). An international conservation community emerged toward the end of the nineteenth century, triggered in part by the establishment of Yellowstone National Park in the United States in 1872. The idea of setting aside pristine areas to preserve them from civilization's pollution and destruction had a powerful appeal in Western thought, expressed, for example, in Romantic and Transcendental philosophies (Muir 1909; Thoreau 1992 [1854]). In the early twentieth century, English preservation organizations had a particularly powerful impact because of Britain's global empire. These groups pressured colonial authorities around the world to establish national parks and then ban native settlements and sharply limit the activities of white hunters in the preserves. They also pushed for rules to protect specific animals and plants judged to be threatened with extinction. This international conservation movement gained strength after World War II, institutionalizing its tenets in influential multilateral organizations including UNESCO and the International Union for Conservation of Nature (IUCN). These institutions, and other independent national conservation groups from the United States and European countries, funded the creation of national parks worldwide.

Efforts at Resolution

In the 1950s, in a nonviolent transition, Tanganyika moved toward independence from Britain. In 1954, Julius Nyerere formed a political party that rapidly attracted strong popular support, and the British authorities acquiesced to the coming changes. Nyerere became prime minister of a self-governing Tanganyika in 1961 and president of the fully independent nation a year later. In 1964, he led a merger between Tanganyika and Zanzibar to form Tanzania, with its current boundaries.

At independence, representatives of the international conservation community, particularly the IUCN and the Africa Wildlife Foundation, urged Nyerere to continue existing conservation policies and maintain the country's network of national parks and nature reserves (Neumann 1998; Goldstein 2005). In return, they offered funding and technical expertise and made the case that the parks were a valuable and enduring source of foreign exchange from hunting and game-viewing tourism. In 1961, Nyerere publicly committed to this conservation program. Even when introducing his plan for a socialist Tanzania in the 1967 Arusha Declaration, Nyerere included a clause reiterating that the nation would hold its natural resources in trust for future generations.

During the socialist period, from the late 1960s through the mid-1980s, the government strengthened the centralized control of conservation and park management it had inherited from the colonial regime. Under this state-centered approach, locals had no input. For example, a new law passed in 1974 clarifying the status of the country's expanding network of reserves referred to local communities only in enumerating activities prohibited to them and outlining punishments for violations. Following suit, the Ngorongoro Conservation Area Authority in the mid-1970s prohibited even subsistence cultivation throughout the NCA and banned grazing in Ngorongoro Crater, the most fertile section of the otherwise arid terrain of the highland region. Emphasizing international concern over the status of Ngorongoro, UNESCO named the crater a World Heritage Site in 1979, the same year that the Everglades National Park was added to the list.

In the mid-1980s, Tanzania abandoned its socialist experiment, which Nyerere himself admitted had failed. At the same time, three other international trends emerged that would affect management of the NCA. First, the rights of indigenous peoples moved higher on the international agenda. In 1982, the United Nations convened a preliminary working group on indigenous rights. In 1995, the General Assembly launched the first International Decade of the World's Indigenous Peoples. By 2002, the UN had established a Permanent Forum on Indigenous Issues. This trend has strengthened the claims of the Maasai for access to land and pasture and for opportunities to participate in decision making.

Second, over the same time period and to some degree in tandem with the international effort to promote indigenous rights, conservationists began to experiment with community-based natural resource management (CBNRM) (Balint 2006). This approach attempts to integrate environmental protection and local community development by devolving authority for natural resource management to the community level and developing opportunities for locals to earn socioeconomic benefits from sustainable use of the resources. In Africa, CBNRM projects often center on regulated safari hunting, game viewing, and cultural tourism on communal lands. Projects typically also include efforts to build local capacity and strengthen local institutions to ensure that the resources are effectively managed and that community benefits are equitably shared (Buck et al. 2001). In principle this approach reduces conflicts between park management authorities and local communities and improves both conservation and development outcomes.

The third major global trend that accelerated in the 1980s and has affected NCA management over the past two decades is the rapid growth in international tourism, particularly nature tourism. International tourist arrivals to Tanzania rose from about 325,000 in 1996 to about 750,000 in 2008 (World Bank 2006). About 40 percent of international tourists to Tanzania visit Ngorongoro Crater, increasing revenue but also escalating pressure on the ecosystem.

Current Situation

In the context of these global trends, tensions in the Ngorongoro Conservation Area have continued. The decision-making dilemma for park management authorities now clearly meets the criteria of a wicked problem. Ngorongoro Crater and the neighboring Serengeti National Park are the main income earners in a national tourism sector that now contributes about one billion dollars annually to the Tanzanian economy, representing about 15 percent of gross domestic product and about 40 percent of foreign exchange. Although Tanzania has worked to emphasize high-end tourism, aiming to maximize earnings while minimizing impacts, tourist overcrowding in the NCA is rising. In competing aggressively for interesting sights, tour operators' vehicles often disturb the wildlife, erode the parklands, and undermine the quality of the game-viewing experience. Even so, developers are building new lodges on the crater rim, reflecting the strong economic pressure to increase the flow of tourists and tourism revenue.

At the same time, the number of Maasai in the conservation area has risen rapidly, from about ten thousand at the time of the NCA's creation in 1959 to over fifty thousand by 2000 (McCabe 2003). Given limited pasture, however, livestock numbers in the NCA—a good measure of Maasai wealth in an area where the government restricts cultivation—have remained largely constant over the past several decades even though the number of people has quintupled. About 60 percent of NCA residents are below the national poverty line, and nearly 40 percent are categorized as very poor or destitute.

The question of cultivation in the NCA remains controversial. In 1975, the Ngorongoro Conservation Area Authority banned all cultivation, but in 1992 the agency moved to permit subsistence cultivation (using handheld hoes only). Then in 2001, Tanzania's prime minister announced that the ban on all cultivation would be reinstated. But only a month later the nation's president overruled the prime minister and stated that small-scale cultivation would again be permitted (McCabe 2003).

The argument over cultivation reflects a deeper conflict over the very presence of the Maasai in the NCA. In the late 1980s the NCAA considered relocating all area residents. In 2002, the authorities threatened to evict all who had arrived after 1975. While the NCAA has not yet followed through on any large-scale resettlement, it often makes the case that effective conservation will ultimately require removing local residents.

Some independent ecological studies, on the other hand, indicate that, as currently practiced, Maasai pastoralism and small-scale cultivation have not adversely affected wildlife and habitat (Homewood et al. 2001; McCabe 2003). Even the attitudes of foreign visitors have a role in the debate. Studies indicate that tourists accept the presence of small plots cultivated by natives in traditional clothing as a valid and picturesque part of the safari experience but that they find the sight of larger-scale, more modern farming inappropriate and discomfiting.

Thus the NCAA faces a contested decision-making environment involving multiple vocal constituencies with strongly divergent values and priorities. The Maasai and the global indigenous rights activists that support them argue for a priority on local rights, participation, and development. At the same time, powerful stakeholders at the national and international level promote conservation of wild animals and habitat. The enormous revenue stream that Ngorongoro generates from tourism intensifies the political pressure to balance these competing interests.

Europe's Emissions Trading System

The European Union's effort to reduce carbon dioxide emissions also has the distinguishing characteristics of a wicked problem. First, people frame, prioritize, and value the underlying issue—global warming—very differently. The British government's chief science advisor, for example, has called it a more dangerous challenge than global terrorism (BBC News 2004). On the other hand, a panel of eminent economists, including several Nobel Prize winners ranked it outside the top ten in global concerns deserving immediate attention, below, for example, improving nutrition, immunizing children, reducing the cost of schooling, educating girls, and widening access to family planning (Copenhagen Consensus 2008). More broadly, a critical source of disagreement, among both experts and nonexperts, is whether—and, if so, to what extent—governments should sacrifice current economic growth to mitigate projected adverse effects of global climate change (Yohe et al. 2008).

Substantial scientific uncertainty helps fuel these divergent views. Experts generally accept two propositions—that global average temperatures are rising and that these changes result at least in part from greenhouse gases added to the atmosphere through human activities (Boko et al. 2007). Yet the answers to central questions about the likely consequences and appropriate policy responses remain uncertain. Timelines and specific outcomes for particular localities are still largely unpredictable. Some regions may actually benefit from projected changes. For example, agricultural yields may increase in some places, Arctic shipping routes are likely to open up, and costs and risks associated with winter cold may decline. Yet there may also be dangerous tipping points beyond which irreversible, highly disruptive changes will occur on a global scale. Scientists cannot yet estimate with confidence whether, and if so when, we may cross such thresholds. Similarly, experts cannot predict how rapidly new technologies may emerge that will allow us to avoid, mitigate, or adapt to adverse changes. Consequently, analysts disagree in estimating the social costs and benefits of various alternative policy responses to climate change (Nordhaus 2007; Stern and Taylor 2007).

Global warming also raises multiple ethical dilemmas, both in the present and the future. Economic and ecological models suggest, for example, that poor countries are likely to suffer disproportionately from global warming, particularly in reduced crop yields and coastal flooding (Boko et al. 2007). Yet rich countries are most responsible for the increased concentrations of greenhouse gases in the atmosphere to date, raising the question of whether rich countries should compensate poor countries for associated harms. Moreover, because of the problem's long time horizons, future generations will face the costs or reap the benefits of choices current governments make over the next few decades. Considerable disagreement remains over how nations and the global community should weigh potential benefits to future generations against the needs of the present.

Policy History

The United Nations first adopted a nonbinding Framework Convention on Climate Change in 1992. The convention set the ambitious but poorly defined goal of stabilizing concentrations of greenhouse gases in the atmosphere at levels low enough to prevent severe disruptions to natural and social systems. In 1997, in Kyoto, Japan, member nations reached agreement on more specific guidelines for achieving this objective. The Kyoto Protocol set targets for industrialized nations to reduce emissions to an average of about 5 percent below their 1990 emissions levels by 2008–12. For many rich countries, this represented a reduction of more than 30 percent from business-as-usual projections for the target period. In 2005, with Russia's ratification, the treaty passed the required threshold of formal support from UN member nations and came into effect.

Despite this political progress, international agreements remain weak. Treaties currently exempt China and India (and the rest of the developing world) from emissions reductions requirements. The United States refused to ratify international accords in part because they do not include these large emerging economies. Thus the United States, China, and India, three influential nations that, combined, produce nearly half of all global emissions, are not yet included in international agreements. Moreover, existing treaties have no enforcement mechanisms even for those nations that are in principle bound by their provisions. In combination, these factors undermine the incentives for signatory nations to meet their obligations and the global impact of any reductions that individual countries may achieve. Even though the 2008–12 target period has arrived, global emissions and atmospheric concentrations of greenhouse gases continue to rise.

Action in the European Union

Going against this tide of slow progress, however, the European Union has developed and implemented a plan to limit its emissions of carbon dioxide, the greenhouse gas that has so far contributed the most to global warming (Ellerman and Joskow 2008). This strong policy commitment, despite the much more ambivalent responses of other countries, follows from the internal politics of EU member nations. Public opinion is conspicuously proenvironmental in much of western and northern Europe, and national leaders have reflected this position (European Commission 2008). While attitudes favoring environmental protection among the public and politicians in the newer EU members of Eastern Europe are generally weaker, these countries adhere to standards set in Brussels.

The European Union's policy is a cap-and-trade regime, known formally as the Emissions Trading System. The program aims to curb emissions across Europe from about twelve thousand major fixed installations, primarily power plants and other large factories. Many analysts favor cap-and-trade for its potential to reduce the total social cost of achieving a given level of emissions reductions (Freeman 2006). Under a cap-and-trade plan for carbon dioxide, the regulatory authority distributes to participating facilities a number of allowances, each permitting the emission of one ton of the gas. The total number of allowances distributed thus constitutes a cap on aggregate carbon dioxide emissions from the facilities during the covered period. The plan reduces emissions by setting the cap below the amount of carbon dioxide that the facilities would emit under business-as-usual practices. Regulators can further reduce the cap in subsequent periods to achieve still lower levels of emissions.

Under a cap-and-trade system, facilities can trade allowances—that is, a market emerges in which facilities can buy and sell emissions permits. In principle, this market mechanism produces an economically efficient outcome. Facilities that can reduce emissions cheaply will do so and then sell unneeded allowances to facilities that have higher marginal abatement costs. Thus facilities that can cut emissions more cheaply will do most of the cutting, and the total cost to society will be minimized.

Consider for example a situation in which two facilities, A and B, are each projected to emit fifty tons of carbon dioxide over a given time period. Assume that under a newly enacted standard the regulatory authority requires the facilities to reduce their total aggregate emissions from one hundred tons to ninety tons. Assume further that the cost of reducing emissions is ten dollars per ton at Facility A and twenty dollars per ton at Facility B. This cost difference may result from the relative age or size of the plants, the availability of alternate fuels, the quality of pollution abatement technologies already installed, or other factors. Under a conventional policy approach, the regulatory agency might require the same level of emissions reductions at each facility. In the case under consideration, an equal-reduction approach would require both facility A and facility B to reduce emissions by five tons. Given the per-unit costs of reducing emissions at the two plants, the total cost of achieving the ten-ton reduction this way would be one hundred fifty dollars (fifty dollars for facility A plus one hundred dollars for facility B).

Under a cap-and-trade regime, the regulator instead would give each facility forty-five allowances, setting the cap at ninety tons of carbon dioxide, and permit the facilities to establish a market for the allowances. In this situation, facility A could choose to reduce its emissions from fifty tons to forty tons, instead of simply reducing to the forty-five tons for which it has allowances. It could then recoup the additional cost of going below forty-five tons—and perhaps some of the cost of its initial five-ton reduction, as well—by selling the five emission allowances it no longer needs to facility B. Facility B could thereby maintain its emissions at fifty tons and avoid incurring its higher per-ton reduction cost. As long as the two facilities negotiated the sale for between ten and twenty dollars per allowance, both facilities would be better off than they would have been under the conventional policy mandating equal reductions. The distribution of costs of compliance between the two firms would depend on the negotiated price of the allowances. But in any event the total social cost of reducing the aggregate emissions to ninety tons—that is, the cost that would ripple out through the economy—would be one hundred dollars (facility A's cost for reducing its emissions by ten tons) rather than the one hundred fifty dollars that would occur under a conventional, nonmarket regulatory regime.

The 1997 Kyoto Protocol explicitly incorporated incentive-based market policies such as cap-and-trade as the best way to slow the buildup of greenhouse gases in an economically efficient way (Ellerman and Joskow 2008). The United States had successful experience with a domestic cap-and-trade program for sulfur dioxide, first implemented in the mid-1990s (Ellerman, Joskow, and Harrison 2003). In an analysis conducted a decade after the cap-and-trade program for sulfur dioxide had been established, economists estimated that the market for emission permits had reduced the total social cost of achieving compliance by as much as 50 percent compared to conventional, nonmarket regulatory approaches (Freeman 2006). The success of this program dramatically reshaped the debate over environmental policy in the United States and elsewhere. Stakeholders across the political spectrum were impressed with the positive outcome and began to look for incentive-based responses to a range of environmental problems.

The European Union implemented its cap-and-trade program for carbon dioxide emissions in 2005. The first three years constituted a trial period. Despite rushed planning and the destabilizing effect of ongoing EU expansion, the program successfully demonstrated that a multinational market for carbon emissions is practical and that industry will respond to market price signals. The initial trading period also provided lessons that the European Union is attempting to use to improve outcomes in subsequent periods. Yet many of the challenges that have arisen are characteristic of wicked problems and will be hard to overcome.

First, because EU member states are sovereign nations, the initial plan gave each nation individual authority to set its own cap, determine which facilities would be covered, and monitor compliance. This contributed to significant inconsistencies across nations and undermined the EU's ability to set an aggregate cap at an appropriate level. For the next round, EU authorities are proposing centralizing these tasks. However, the wide variety in economic structure and industrialization across nations creates conflicts and uncertainties. The recent East European entrants have joined legal action to challenge the proposed allocation process for the second trading period, arguing that new caps proposed on their industries are too strict and ignore local economic conditions.

A second key problem is that public support for emissions reductions dropped sharply once the price increases linked to the imposition of the cap spread through the economies. Attitudinal polls and behavioral surveys reveal that Europeans, like citizens of other countries, often have stronger support for environmental sacrifice in principle than in practice. This disconnect emerged once electricity generators began to pass on the costs of the allowances to consumers.

Consumers' perceptions of windfall profits for power companies exacerbated public dissatisfaction. The question of how to best allocate allowances when beginning a cap-and-trade plan is a puzzle for policy makers (Ellerman and Joskow 2008). Distributing allowances to emitters for free, as happened in the first round of the EU's Emissions Trading System, has the benefit of reducing industry resistance to emissions reductions. This method limits the penalty that would otherwise be imposed on industry for strategic business decisions made under prior policy regimes in which carbon dioxide emissions were not charged.

Auctioning allowances, on the other hand, has the benefit of requiring firms to pay the market price for valuable assets that, in effect, can be thought of as belonging to the public at large—that is, permits to add greenhouse gases to the public's atmosphere. An auction also raises revenue for other useful social purposes related to the imposition of the cap-and-trade plan. These may include investments in carbon-free energy sources or relief for low-income consumers likely to suffer disproportionately from electricity price increases.

In principle, the efficiency gains from a market-based regulatory policy will occur whether the value of the new assets—allowances—initially accrues to industry, through free allocation, or to the government, though auctions. But this policy choice inevitably engenders strong political battles over how the value of these new assets is distributed.

The scale of the problem, encompassing twenty-seven nations with divergent economic and political regimes, adds to the complications. Some EU countries have highly regulated electricity markets, while others have moved toward deregulation. The effects of free or auctioned allowances will vary significantly across these economic systems, making it difficult for central authorities to promote economic efficiency and manage political dissatisfaction.

A third fundamental problem with the cap-and-trade program relates to determining the appropriate level for the cap. Given scientific, technological, and economic uncertainty, policy makers cannot determine at the outset what cap, or trajectory of cap reductions over time, will achieve the desired outcome of keeping atmospheric greenhouse gas concentrations below the level that will trigger severe global social and ecological disruptions over the long term. Nor can policy makers estimate with confidence what levels of emissions reductions can be achieved without triggering severe political and economic disruptions over the shorter term. Firms, and ultimately consumers, are likely to push for higher caps to reduce short-term costs.

Current Situation

In undertaking to reduce its greenhouse gas emissions, the European Union faces the key defining characteristics of a wicked problem: multiple stakeholder groups with divergent value systems, priorities, and incentives; policy makers facing multiple scientific, technical, and economic uncertainties; and a complex, large-scale, policy-making environment with an extended time horizon.

Moreover, the political context in this case is particularly challenging. The European Union is a multinational confederation that requires unanimous agreement from twenty-seven diverse nations to adopt the administrative streamlining necessary to manage its climate change policy—and many other policy problems—more effectively. Finally, the European Union is attempting to achieve global goals without global support.

The Sierra Nevada

California's Sierra Nevada stretches approximately 450 miles through the northeastern part of the state from near the Oregon border to the southern end of the Central Valley. This mountainous region has features of exceptional natural beauty, including Yosemite Valley, Lake Tahoe, and many rugged summits above ten thousand feet, including Mt. Whitney, the highest peak in the lower forty-eight states.

Much of the terrain is forested, with diverse forest structure reflecting the region's variations in topography, soils, and climate (Barbour and Majors 1990). Groves of thousand-year-old giant sequoia on the western slopes differ significantly from pine-dominated, open park-like stands on the range's arid east side. Even within the mixed conifer forests common on the western side, forest composition and structure vary with altitude, latitude, and aspect (Davis and Stoms 1996). These different forest zones also have different histories, with most of the effects of modern human settlement, livestock grazing, and timber harvest found at lower elevations (McKelvey and Johnson 1992). The variety of forest habitat supports a wide range of animal life. In all, scientists have identified more than 550 vertebrate species in the Sierra Nevada (USDA Forest Service 2001a).

Climate helps shape the region's ecology. Prevailing winds carry moisture off the Pacific Ocean up the slopes of the range, leading to substantial snowfall at higher elevations on the western side and dry conditions to the east of the crest. The snow pack on the higher western slopes commonly reaches several feet in depth and feeds streams and rivers that flow year-round despite limited rainfall. Given the limited precipitation at lower elevations on the western side and at all elevations on the eastern side, fire has played a central role in the natural ecology of the region (Skinner and Chang 1996).

Archeological evidence indicates a human presence in the Sierra Nevada for thousands of years (Duane 1996; Hull 2007). Prior to European settlement, the combination of dry summers, lightning, and fire use by Native Americans resulted in a typical fire regime of low-intensity surface or ground fires recurring on average every decade or two. Occasionally, the coincidence of local fuel accumulations and particularly hot dry weather would generate stronger crown fires that created openings in the forest canopy of perhaps three to five acres where regeneration of shade-intolerant species such as pines and giant sequoia could occur.

The first European references to the Sierra Nevada—descriptions of snowcapped peaks seen in the distance—came from Spanish explorers and missionaries in the late 1700s (Farquhar 1965). At this time, inland California was still a remote, largely unexplored hinterland of the Spanish colony in Mexico. Between 1820 and 1848, the territory underwent a series of major political shifts. Mexico achieved independence from Spain in 1821. Over the next two decades Mexican cattle ranchers of Spanish descent established properties in southern California. Over the same period, English-speaking hunters, trappers, and settlers began to move into the region from the north and east. In 1846, the English-speaking settlers declared their independence from Mexico. This rebellion was quickly overshadowed by the larger Mexican-American War of 1846–48. In the peace treaty of 1848, Mexico ceded control of California and much of what is now the southwestern United States. While Arizona and New Mexico remained territories until the early twentieth century, California became a state in 1850.

The discovery of gold helped drive the rapid transition to statehood. At the time of the end of the Mexican-American War, California was still lightly settled. In 1848, however, a settler found gold on the western slopes of the central Sierra Nevada. By 1849, once publicity surrounding the discovery spread to the eastern United States, Europe, and beyond, a surge in population began that increased California's population by a factor of ten. Many of the new arrivals moved to mining towns along the western side of the Sierra Nevada. This rapid growth drove broader, longer-term economic development both along the coast and in the interior. Networks of roads, railroads, and shipping lines spread and commercial agriculture, livestock husbandry, and timber harvesting expanded.

Repercussions of Development

The impact of agriculture, livestock, and logging in the Sierra Nevada ultimately led to a battle over land use that would have implications for environmental management that are still relevant today. The concepts of *preservation* and *conservation*—two views regarding the relationship between humans and nature, and the stewardship responsibilities that humans have for nature—gained strength toward the end of the nineteenth century (Nash 1989). Preservation in this sense has links to the transcendentalist movement of Emerson and Thoreau. The preservation ethos advocates setting aside special natural areas for recreation and *spiritual nourishment*, and excluding any commercial extraction and exploitation within their boundaries (Nash 1989). Conservation, on the other hand, which fits within the framework of utilitarian and progressive philosophies, advocates sustainable and efficient use of natural resources following scientific principles to promote the general well-being of society at large.

In the United States at the turn of the twentieth century, these contrasting views were to considerable degree personified in John Muir and Gifford Pinchot (Nash 1989; Meyer 1997) and the two distinct policy approaches they advocated for the Sierra Nevada. Echoes of this difference contribute to the *wickedness* of all four of the environmental management dilemmas we discuss in this book, but they can be heard most clearly in the continuing conflict over forest management in the Sierra Nevada.

John Muir was an avid amateur naturalist who became a charismatic spokesman for wilderness preservation. His activism matured in the Sierra Nevada. He first visited the Yosemite Valley in 1868 at age thirty and was entranced by its physical beauty and spiritual power. Over the next forty years he fought to protect wild nature in the Sierra Nevada and beyond. He founded the Sierra Club; wrote influential books and articles promoting the preservationist ethic; and successfully lobbied President Theodore Roosevelt to take Yosemite under federal control, setting the stage for its formal designation as a national park. In his later years, he ultimately lost a bitter battle over building a dam on the Hetch Hetchy River in the northern end of the park to create a reservoir for San Francisco.

Gifford Pinchot, with an elite education, formal graduate training in forestry, and a strong dedication to public service, was at the vanguard of national forest policy formulation at the turn of the twentieth century. As the first chief of the Forest Service, he professionalized the agency, instilling methods of scientific management as the means for achieving the goal of maximum contribution to the social good. As a protégé of Theodore Roosevelt, his fellow Progressive, he was able to implement his ambitious national vision for efficient use of public lands.

In the early 1890s, Muir and Pinchot saw themselves as allies. Both decried the wasteful use of natural resources and the tendency of many to aim for narrow, short-term gain without considering long-term outcomes. Their friendship ended, however, as they came to disagree first over sheep grazing in the Sierra Nevada and later over the plan to dam the Hetch Hetchy in Yosemite. The well-publicized split between these two leaders divided those concerned about the environment in lasting ways. To some degree the approaches to public land management embodied in national parks, protected from extractive activities, and national forests, managed for multiple uses, reflect these divergent visions.

The federal government initially established national forests in the western United States under the Creative Act of 1891, which permitted the president to withdraw lands from the public domain as forest reserves. These forest reserves came under the jurisdiction of the Government Land Office in the Department of the Interior until 1905, when they were transferred to the Forest Service in the Department of Agriculture (USDA Forest Service 1993).

A letter dated February 1, 1905, from the secretary of agriculture to Gifford Pinchot, the new chief of the Forest Service, explicitly articulated the progressive and utilitarian ideals of sustainability and efficiency in public land management: "all land is to be devoted to its most productive use for the permanent good of the whole people and not for the temporary benefit of individuals or companies. All the resources of forest reserves are for use....under such restrictions only as will insure the permanence of these resources....In the management of each reserve local questions will be decided upon local grounds...and where conflicting interests must be reconciled, the question will always be decided from the standpoint of the greatest good of the greatest number in the long run" (USDA Forest Service 1993). Although the letter appeared under the secretary's signature,

Pinchot himself had prepared this seminal document, laying out the goals of the Forest Service (Pinchot 1947).

For the next seventy-five years, the Forest Service's stewardship of the national forests aimed to achieve the utilitarian goals of meeting the people's needs for wood, water, forage, and economic development (Nelson 1999). At the beginning of World War II, the need for wood to support the war effort resulted in large increases in timber harvests in the national forests. Following the war, harvesting continued at an accelerated rate to satisfy the demand for low-cost housing.

These forest management policies, including fire suppression and the expansion of timber harvesting, produced important social benefits and generally received broad public and congressional support. But these policies also laid the foundation in unanticipated ways for current policy dilemmas. The public consensus favoring extensive timber harvesting and other extractive activities began to fracture in the last several decades of the twentieth century, driven on the national level by the emergence of the modern environmental movement and on the regional level by the substantial demographic shifts that occurred in the Sierra Nevada.

Until the 1970s, local economies in the region generally depended on the use of natural resources in the neighboring forests. But the population has grown rapidly in recent decades, leading to important social and economic changes. During the 1990s, the region's population grew by about 25 percent, compared to about 17 percent for California as a whole during the same period. The counties of the Sierra Nevada are now home to about 3.8 million people. This population growth has stressed the counties' infrastructures and sharply increased demands on water and other environmental services.

This large influx of people, often with different perspectives and expectations from longer-term residents, coupled with the expansion of residential and commercial development, has led to both shifts in public attitudes toward the forests and increased risk to lives and property from wildfire. In conjunction with national trends toward increased environmental activism and litigation, these changes have combined to undermine long-standing support for the Forest Service's traditional utilitarian goals and practices. The ongoing gridlock over the Sierra Nevada Forest Plan Amendment is in large part a consequence of these broad cultural and demographic shifts. In the next paragraphs, we outline the history of the conflict.

The Sierra Nevada includes ten national forests and one administrative unit encompassing 11.6 million acres. Administrative responsibility for the region is vested in the regional forester for the Pacific Southwest Region

(Forest Service Region 5). A forest supervisor heads each forest within the region. Each forest is further subdivided into smaller administrative units, called districts or ranger districts, each supervised by a district ranger. Historically, the Forest Service has followed a decentralized administrative structure with significant authority delegated *close-to-the-ground*. Of the agency's approximately thirty-five thousand employees, fewer than a thousand are based in the national headquarters in Washington, DC.

The agency operates under various federal statutes, including the Multiple Use-Sustained Yield Act of 1960, the National Environmental Policy Act of 1970, the Endangered Species Act of 1973, and the National Forest Management Act of 1976. These acts sometimes set unrealistic goals and sometimes conflict with each other. Beginning with the National Environmental Policy Act, environmental laws generally require public participation and permit citizen lawsuits.

Both the Endangered Species Act and the National Forest Management Act have especially rich litigious histories. Forest plans and individual projects, particularly those involving timber sales, are often appealed and then litigated. This necessitates the preparation of lengthy and complex planning and decision documents to meet court requirements for sufficiency. The process to amend a forest plan often requires five to ten years or more. The agency's effort to clarify and streamline national forest planning has itself continued without resolution for over fifteen years through two presidential administrations and two rounds of litigation, while generating tens of thousands of public comments.

The Sierra Nevada Forest Plan Amendment process began in the early 1990s following a contentious battle in the Pacific Northwest Region (Forest Service Region 6) over northern spotted owls. The controversy centered on whether Region 6 was preserving sufficient old forest habitat to protect the owl. A forest plan implemented in 1986 had reduced but not eliminated timber harvesting in old-growth forests. Environmental groups saw an opportunity to protect large areas of old-growth forests along with the northern spotted owls by forcing the Forest Service to follow language in regulations associated with the National Forest Management Act that requires the agency to provide for "diversity of plant and animal communities"; "preserve the diversity of tree species"; and maintain "viable populations" of all species throughout their range. The Audubon Society and the Sierra Club brought the case to court. Complicating matters further, the US Fish and Wildlife Service in 1990 formally listed the northern spotted owl as "threatened" under the Endangered Species Act. Such listing requires protection of habitat. The acrimonious, polarizing debate over the spotted owl in the Pacific Northwest had the effect of energizing and mobilizing many competing interest groups, both locally and nationally.

In 1991, a federal court ruled that Region 6's strategy to protect northern spotted owl habitat was insufficient to meet regulatory requirements. The court blocked further timber sales until a satisfactory plan could be developed. In conjunction with the broader social changes described earlier, this decision helped drive a sharp downward trend in timber harvests that occurred in the early 1990s.

The court's ruling on the inadequacy of the plan to protect the northern spotted owl in the Pacific Northwest also called into question a similar strategy for a related species, the California spotted owl, found in the Sierra Nevada. In response, the regional forester in the Sierra Nevada at the time (Ron Stewart, one of the authors of this book) commissioned a scientific assessment. The report based on that assessment, issued in 1992, found that the strategy in Region 5 was indeed insufficient (Verner et al. 1992). The same report, however, also recognized a significant risk to owl habitat from catastrophic wildfire and recommended an interim strategy of aggressive forest thinning from below to reduce understory fuels.

In hindsight, these findings appear to presage the full complexity and divisiveness of the policy dilemma that was emerging. For example, both opponents and proponents of timber harvesting and active forest management could claim based on this independent scientific assessment that their favored policy in some form could help protect the owls. Moreover, as indicated in the contentious debate in the Pacific Northwest, the specific questions concerning spotted owls and old-growth habitat were evolving into a much broader struggle over the proper role of the Forest Service in natural resource management. In the new political environment, the agency could no longer pursue its traditional utilitarian goals with confidence that scientific principles would provide sufficient guidance and that the general public would respect the agency's expertise. To some degree, the earlier conflict between followers of Muir and Pinchot was reemerging in a new guise.

Nevertheless, in 1993, hoping to avoid the intervention of the courts as had occurred in the Pacific Northwest, Region 5 implemented the interim strategy of forest thinning and fuels reduction recommended in the scientific assessment, and set a two-year timeline for developing an environmental impact statement for a new long-term forest plan. In working to complete the environmental impact statement, the region undertook an unprecedented effort to marshal the best science and engage the full range of public stakeholders. Despite the agency's best efforts from 1993 to the present, however, the conflict over policy alternatives continues. As we discuss in later chapters on our research on the Sierra Nevada Forest Plan Amendment case, the key areas of contention have been the interrelated issues of whether to harvest timber (and if so, how, how much, and where); how best to protect habitat and species; how best to reduce fuel loads in the forest to limit the risk from wildfire; and how to pay for forest management when the era of large-scale, revenue-producing logging appears to be over.

Current Situation

There have been many twists, turns, and detours in the struggle to achieve a broadly acceptable plan for the Sierra Nevada national forests. These include various scientific studies, numerous meetings with stakeholders, thousands of public comments, multiple environmental impact statements, several congressional interventions, and two separate records of decision. The protracted planning effort in the Sierra Nevada has now spanned three presidential administrations, six regional foresters, and five chiefs of the Forest Service. Ultimately the region was unable to avoid the intervention of the courts. In May 2008, for example, a court ruling blocked logging of three tracts for fuels management in the northern Sierra Nevada (Egelko 2008).

With a grant from Region 5, we conducted research on the Sierra Nevada Forest Plan Amendment dilemma in 2003–4. As part of this project, we undertook a wide-ranging review of the literature in many disciplines. During fieldwork, we elicited stakeholder attitudes and preferences. Later we analyzed the data we had collected to explore the conflicting views from new angles. Our research on Sierra Nevada Forest Plan Amendment led us to the notion of wicked environmental problems and the various responses that have been attempted. In chapters 6 through 9, we discuss in greater detail the Sierra Nevada decision process and the findings from our research on this important case study.

Conclusions

The examples we describe in this chapter offer variations on the theme of wicked environmental problems. All four cases involve management dilemmas extending over large scales and long time horizons. In all four cases, multiple and compounding risks and uncertainties combine with sharply divergent public values to generate contentious political stalemates. No optimal solutions exist; yet policy makers must act. Throughout the book, we refer to these cases to illustrate our points regarding the strengths and limitations of possible responses to wicked problems, including the precautionary principle, adaptive management, and public participation. We return to these examples again in later chapters as we introduce and discuss our proposal for a *learning network* approach, enhanced through systematic elicitation and analysis of public preferences.

Chapter 4

The Precautionary Principle

Wicked environmental problems are characterized by scientific uncertainty, deep public disagreement over desired states and preferred outcomes, the impossibility of finding an optimal solution, and the requirement that despite these unknowns and conflicts the responsible decision maker must act (Allen and Gould 1986). In these conditions, public managers—whether or not they recognize that they face a wicked problem—often respond by applying such strategies as the precautionary principle, adaptive management, or public participation. Yet these approaches, whether used singly or in combination, have generally not helped manage the wickedness of the problems. To lay a foundation for our discussion in chapters 7–9 of an enhanced learning network process that may be an improvement over existing approaches, we consider the three more commonly used responses in turn: the precautionary principle in this chapter, adaptive management in chapter 5, and public participation in chapter 6.

In this chapter, we begin with a review of the history of the emergence and evolution of the precautionary principle, a discussion of its general strengths and limitations as a basis for policy making, and a consideration of proposed modifications to strengthen its practical applicability. Then, to give context to the points made in this general overview, we offer a detailed analysis of the influence of the precautionary principle in the particular case of the Sierra Nevada Forest Plan Amendment (SNFPA) process.

Overview

Although the precautionary principle is described in various ways in the scholarly literature (O'Riordan and Jordan 1995; Manson 2002; Cooney and Dickson 2005a), its basic concept is captured in the adage "better safe than sorry" (Pielke 2002). The key idea is that technologies or processes with the potential to harm human health and the environment should be regulated, even if the nature and likelihood of the potential harms that may result are uncertain. Traditionally, societies have tended to give the benefit of the doubt to technologies and processes that contribute to economic development. Associated harms, such as health hazards, workplace dangers, or environmental degradation, are typically addressed after the damage has become unmistakably apparent and countries have become rich enough to accept the costs of mitigating these adverse outcomes.

The precautionary principle articulates an alternative philosophy, asserting that potential long-term, adverse, unintended consequences should be considered in advance rather than addressed after the fact. In other words, the precautionary principle shifts the burden of proof. Under common business-as-usual conditions, those concerned about the potential adverse effects of current or proposed technologies and processes must provide definitive evidence of future danger. Conversely, under the precautionary principle, proponents of these technologies and processes have to provide evidence that they will be safe over the long term. Under the precautionary principle, a lack of clear evidence that current or proposed practices will cause future harms cannot serve as a justification for delaying action to regulate them (Raffensperger and Tickner 1999).

To give a current example, the debate over global warming frequently includes arguments that implicitly favor business-as-usual over precautionary actions, or vice versa. In arguing against efforts to limit greenhouse gas emissions, those opposing regulation frequently point to scientific uncertainty regarding both the extent of human contributions to climate change and the severity of the adverse effects that may occur. They generally call for further research to reduce the uncertainties before costly emission-reduction policies are implemented. Proponents of precautionary action to counter climate change, on the other hand, argue that the likely adverse effects of a buildup of greenhouse gases in the atmosphere are serious enough to justify potentially costly regulation despite remaining uncertainties. Similar arguments and counterarguments are commonly expressed in debates over the health and environmental risks posed by genetically modified foods, trace amounts of pharmaceuticals in drinking water, biodiversity loss, and various other issues. The relative strength of precautionary arguments tends to rise with the perceived severity of possible future harms. In cases where long-term outcomes are potentially catastrophic, precautionary arguments may be seen to justify substantial short-term sacrifices to avoid uncertain but potentially devastating future hazards (Stern 2007). Thus, many debates over policies that promise clear gains in the short term but may add to risks of long-term adverse outcomes often center on claims and counterclaims over the likelihood and severity of the uncertain future harms.

Emergence and Evolution

The precautionary principle first began to influence environmental policy on both sides of the Atlantic in the 1960s. In the United States, the wave of major, ground-breaking statutes relating to human health and the environment adopted in the 1970s include precautionary ideas (Jordan and O'Riordan 1999). For example, the National Environmental Policy Act enacted in 1969 requires that before any proposed government action that may affect the environment can be implemented, agencies must complete environmental impact assessments and consider alternative policies. The Occupational Safety and Health Act enacted in 1970 requires that employers limit workers' exposure to substances suspected but not yet proven to be health risks. The Clean Air Act of 1970, the Clean Water Act of 1972, and the Endangered Species Act of 1973 all include precautionary language mandating *margins of safety* in human health and environmental protection.

In Europe, the precautionary principle can be traced to the term *Vorsorgeprinzip*, coined in Germany in the early 1970s (von Moltke 1988; O'Riordan and Jordan 1995). The term, which translates literally as "principle of advance caring," was first applied in response to observed deterioration in the ecological health of German forests. Acid deposition associated with sulfur and nitrogen emissions from industrial, commercial, and transportation sources was the suspected culprit, but scientists could not provide firm evidence of cause and effect. Despite this scientific uncertainty and the costs of the new policy to businesses and consumers, the German government, with substantial popular support, instituted regulations to reduce power plant emissions linked to acid rain.

After becoming a key component of German environmental law, the concept of precautionary action has gained wide acceptance across Europe and has also been explicitly incorporated into various international treaties and conventions (Dethlefsen 1993; Raffensperger and Tickner 1999). Principle 15 of the Rio Declaration includes the following model language:

> In order to protect the environment, the precautionary approach shall be widely applied by States according to their capabilities. Where there are threats of serious or irreversible damage, lack of full scientific certainty shall not be used as a reason for postponing cost-effective measures to prevent environmental degradation. (United Nations 1992)

Similar statements can be found in many international documents.

Over the past two decades, however, attitudes in Europe and the United States toward the precautionary approach have diverged. In comparison with western European nations and UN institutions, the United States, particularly from the early 1980s on, has been less sympathetic to the precautionary principle. Indeed, there has been considerable resistance to the idea both in Congress and the executive branch. Climate change again serves as a useful example. In 1997, the US Senate passed a resolution on a vote of 95–0 indicating that it would refuse to even consider the Kyoto Protocol for ratification if submitted by President Clinton. The Senate took this action in part because the potential long-term harms the treaty aimed to address were uncertain while the short-term costs to the American economy appeared both certain and immediate. The US government continued its opposition to the Kyoto agreement for the next decade. In contrast, many European countries rapidly ratified the Kyoto Protocol, and, as discussed in chapter 3, the EU has since implemented substantial mandatory policies to curb greenhouse gas emissions.

In general, current skepticism about the precautionary principle in the United States tends to be top down, driven by government officials, business leaders, and agency experts. In Europe, in contrast, the acceptance and application of precautionary thinking tends to be bottom up, driven by voters and consumers. Yet precautionary thinking can be influential in the United States as well, particularly in the context of wicked problems where public activism plays an important role. Later in this chapter, we discuss in detail the influence of the precautionary principle in the Sierra Nevada forest planning process.

Strengths, Limitations, and Possible Variations

Assessments of the precautionary principle variously describe it as a powerful tool for protecting human health and the environment under conditions of uncertainty (Cameron and Aboucher 1991; Dethlefsen 1993; Raffensperger and Tickner 1999; deFur and Kaszuba 2002) or as a poorly defined and unscientific approach with limited value in real-world policy dilemmas (Bodansky 1991; Manson 2002; Pielke 2002; Sunstein 2003). If advocates are right, the precautionary principle can help manage the scientific uncertainty inherent in wicked problems by promoting policy choices that minimize the risk of adverse or catastrophic outcomes, while at the same time reducing political conflicts over management alternatives by providing a broadly acceptable foundation for assessing proposed policies. If critics are right, the precautionary principle may actually do more harm than good on both counts. In terms of policy choices, critics argue that the precautionary principle tends to bias the decision-making process against flexible, adaptive responses and, depending on circumstances, may be just as likely to lead to catastrophic outcomes as other decision criteria. In the political realm, critics contend that the precautionary principle has become a polarizing approach, rather than a plausible unifying position. In the next section, we review arguments for and against the precautionary principle, compare it to alternative policy analysis tools, and discuss modifications that have been proposed to strengthen it in response to criticisms.

Potential Strengths and Advantages over Current Alternatives

Proponents of the precautionary principle argue that it promotes a prudent safety-first attitude toward activities that may pose significant risks to human health and the environment and highlights risks associated with uncertainty downplayed by more conventional decision processes, such as cost-benefit analysis and risk analysis (Raffensperger and Tickner 1999). In mandating a safety-first approach, strong versions of the precautionary principle would entirely ban activities that have the potential to harm human health or the environment even if the scientific case for cause and effect is uncertain. Weaker versions call at minimum for not rushing ahead with new technologies or processes until risks and options have been carefully considered. No matter the formulation, application of the precautionary principle makes it significantly more difficult to introduce or continue potentially harmful activities, notwithstanding their potential benefits.

Advocates also assert that the precautionary principle offers improvement over other tools for evaluating alternative policies under uncertainty. Cost-benefit analysis, for example, a widely used approach for assessing policy options, is based on the assumption that all relevant costs and benefits, both present and future, can be monetized, summed, and compared. In practice, this assumption can rarely be satisfied. Critics of cost-benefit analysis note in addition that the practice of discounting future costs and benefits tends to bias the results of the analysis in favor of policies predicted to generate short-term gains, even if they also threaten long-term harms to human health or the environment. Discounting future costs and benefits also tends to create a bias against actions, such as reducing greenhouse gas emissions or reducing timber harvests in national forests, likely to produce short- to medium-term economic costs but with the potential to generate large future environmental and social benefits (Stern 2007).

In this view, risk analysis is subject to analogous weaknesses. Risk analysis assumes, for example, that hazards and exposure levels are known or can reasonably be estimated and that probabilities of adverse outcomes are understood well enough to support effective relative risk comparisons. Yet sufficient baseline information is difficult to collect in the case of toxic substances or carcinogens, and may be next to impossible to collect for complex environmental problems in which ecosystem responses and socioeconomic feedback mechanisms are poorly understood or fundamentally uncertain.

Further, as discussed in chapter 2, the process of defining risk is to a considerable degree a function of political influence and competition among divergent values and worldviews (Auberson-Huang 2002). Risk assessment includes the steps of calculating the likelihood and potential seriousness of particular outcomes, but it also includes implicit valuations of the importance of the social or ecological characteristics that may be affected. In matters of human health, this process may not be controversial: human life and health are generally accepted to be of primary importance. In complex environmental dilemmas, on the other hand, there may be multiple competing interests and no general consensus. Indeed the lack of a consensus definition of desired outcomes is a defining characteristic of wicked problems, and is common to all four of the cases we introduce in chapter 3.

In the case of the Everglades, for example, participants disagree on how water should be allocated between the urban areas and the wetlands. In Ngorongoro, Tanzania, participants differ on how grazing land should be apportioned between domestic livestock and wild animals. In the EU emissions trading scheme, various interest groups have conflicting priorities regarding how costs and benefits of climate change mitigation should be divided between current and future generations. In the Sierra Nevada, stakeholders disagree over how ecological risks to spotted owls should be balanced against economic risks to local human communities that depend on timber harvested from potential owl habitat. Standard cost-benefit and risk analyses are poorly equipped to address policy options within such value-laden controversies.

While cost-benefit and risk analysis aim to estimate the future value and efficacy of policy alternatives, the precautionary principle is based on a fundamentally different perspective. Advocates of the precautionary principle propose a strategy of "backcasting" as opposed to forecasting (Raffensperger and Tickner 1999). In the precautionary view, instead of haphazardly introducing new ways of doing things that should be—but typically are not—effectively evaluated to determine whether their inherent risks are tolerable, society should instead direct its energies to developing technologies, products, and processes that help achieve broadly beneficial goals. Thus the precautionary approach encourages holistic, long-term thinking about how to reduce risk proactively, instead of following the commonly applied current strategy of attempting to manage risk reactively on a case-by-case basis. Proponents further contend that applying the precautionary principle promotes broader public engagement than is currently the norm in environmental policy making. They argue that this is a positive development because experts alone cannot determine appropriate policies in the context of complex cases with public disagreement about desired outcomes and uncertainty regarding the effects of policy alternatives.

In sum, proponents of the precautionary principle point to several key strengths for the precautionary principle: that it promotes a prudent safety-first attitude, that it highlights long-term risks associated with uncertainty downplayed by more conventional decision processes, such as cost-benefit analysis and risk analysis, and that it encourages broader public engagement in policy making (Raffensperger and Tickner 1999).

Inherent Limitations and Possible Variations

Despite these arguments in favor of the precautionary principle, the approach is also subject to strong criticism (Keeney and von Winterfeldt 2001; Starr 2003; Sunstein 2003). Critics raise the following concerns. The precautionary principle is inconsistently defined and interpreted, may not account for potential risks arising from precautionary actions themselves, and—perhaps most important—typically fails to provide useful concrete guidance to decision makers in particular cases. Critics argue that the precautionary principle's weakness as a practical policy tool appears to stem in part from its implicit incorporation of inherent irrationalities found in common heuristics (Sunstein 2003). Examples of these include the tendency to hold on to what we have even if what we may gain by giving it up is more valuable, the tendency to focus on potential dramatic negative outcomes rather than on the very low probabilities of these outcomes actually occurring, and the tendency to ignore trade-offs and ripple effects—that is, to focus on a particular issue of concern and overlook wider implications. In this view, while cost-benefit and risk analysis may understate important questions of values and downplay possible long-term adverse effects, the precautionary principle tends to underestimate trade-offs and limit management flexibility in the field. Our discussion of the effects of precautionary thinking in the Sierra Nevada case presented in the next section of this chapter provides concrete illustrations of these points.

Several alternative, middle-ground, or hybrid approaches have been proposed in response to criticisms of the precautionary principle. One is the "maximin" principle, which calls for choosing the policy alternative with the least-bad, worst-case scenario (Comba, Martuzzi, and Botti 2002). A second is "risk-risk" analysis, which advocates assessing all risks throughout the system, including those resulting from trade-offs and ripple effects, and then selecting the policy alternative that minimizes aggregate net risk (Goklany 2001). A third is to avoid the "dichotomy trap"—that is, to incorporate the strengths of both scientific and precautionary approaches (Stirling 1999; Peterson 2002).

None of these appears to overcome the limitations, however. The first, based on the principle of catastrophe aversion, suffers from some of the same weaknesses as the precautionary principle. It ignores the probabilities that the worst-case scenarios will occur. In other words, it may lead to eliminating from consideration useful policies that are highly likely to lead to beneficial outcomes because they may have a remote risk of a catastrophic result. Also, this alternative assumes no scientific uncertainty or ignorance. That is, it does not address the possibility that a policy chosen based on the catastrophe-avoidance criterion may lead to an entirely unforeseen catastrophe.

The second proposal is closer to conventional risk assessment and cost-benefit analysis and incorporates several of the weaknesses of these approaches identified above. For example, it assumes that all risk can be identified, agreed upon, quantified, and compared. Significantly, this information burden is likely to increase rather than decrease as public environmental managers, following current best practices, attempt to address problems at larger scales. The third proposed alternative aims to defuse the debate over the precautionary principle by integrating precaution and science. This implies an acknowledgment that, while precautionary thinking deserves a place in policy debates, the precautionary principle by itself cannot provide clear guidance in complex policy dilemmas. Reaching a similar conclusion, a recent international review of challenges associated with the use of the precautionary principle in biodiversity protection proposed a set of guidelines for combining approaches (Cooney and Dickson 2005a). The guidelines call for establishing a policy decision framework that, among other things, incorporates the best available technical and scientific information; identifies and assesses costs, benefits, risks, and uncertainties; incorporates precautionary measures in a manner appropriate to the threats; applies adaptive management, with effective monitoring to provide timely feedback; and encourages the participation of all stakeholders. Yet these guidelines highlight the dilemma involved in attempting to apply the precautionary principle in the policy arena. To be useful it must fit into a framework that includes other approaches with which it may conflict, such as cost-benefit analysis and adaptive management.

The difficulties in operationalizing the precautionary principle arise at least in part because it is, by definition, a principle, not a concrete tool for policy analysis and decision support. As a principle, it provides an important reminder that policy makers should consider and attempt to avoid potential, long-term adverse consequences. But, in the context of wicked problems, where there is no agreement on preferred outcomes and all actions have uncertain consequences, the precautionary principle cannot help identify the least risky policy course.

The Precautionary Principle in the SNFPA Process

The SNFPA process provides a useful example of the implications of the precautionary principle for policy making in a particular case. In 2003, at the beginning of our research into the conflict over forest planning in the Sierra Nevada, Forest Service administrators in the region asked us to examine the role of the precautionary principle. They believed that the strong precautionary attitudes of key stakeholder groups were making it difficult for the agency to address risks and uncertainties in the planning dilemma. These stakeholder groups advocated for policies with lower levels of active forest management on the grounds that active management, including for example forest thinning, threatened old-growth habitat, which they believed was best preserved through the operation of natural processes with minimal human interference.

As part of our research we reviewed the record of public engagement in the decision process and examined the formal documents that emerged, focusing specifically on the 2001 final environmental impact statement and record of decision and the 2003 report of the team that later reviewed the decision for the regional forester (USDA Forest Service 2001a, 2001b, 2003). We also discussed SNFPA processes and outcomes with a range of stakeholders during meetings in the Sierra Nevada region. While the 2001 environmental impact statement and record of decision do not explicitly name the precautionary principle, we found that the precautionary principle's influence is apparent, both in public input and in agency responses.

Among its priorities in the Sierra Nevada, the Forest Service aims to achieve two key management goals: reducing risk from wildfire and protecting old-growth forest habitat. Yet efforts to achieve these two goals can create conflicting demands. That is, strategies designed to reduce risk in one of these priority areas may increase risk in the other. The environmental impact statement acknowledged this dilemma in its discussion of potential conflicts between fuel treatment strategies and protection of forest areas where California spotted owls are present. (As discussed in more detail in chapter 9, fuel treatment in this context refers to efforts to lower the risk of severe wildfires by reducing the amount of combustible material in the forest, typically through mechanical clearing or controlled burning. Contentious debates over the type and location of fuel treatment activities have been important sources of continuing conflict in the SNFPA process.)

As required by law, the 2001 environmental impact statement included analysis of multiple policy alternatives. As a result of participatory processes during development of the statement, these alternatives reflected the preferences of various influential stakeholder groups. One alternative, which responded to concerns articulated by the US Fish and Wildlife Service, the agency responsible for endangered species protection, incorporated a cautious approach to fuel treatment in areas where owls are present, even near human communities, while recognizing that the risk of fire in these areas may be increased as a result (USDA Forest Service 2001a).

Following the logic of the precautionary principle, this alternative would have required proponents of fuel treatment to prove that owls would not be harmed before treatment activities could proceed. As Forest Service managers pointed out, however, precautionary logic would have supported the opposite policy if the starting point were changed. If reducing fire risk were given priority, the precautionary principle would have supported continuing fuel treatment, even in old-growth forest stands, unless opponents could prove that owls would be harmed. In other words, the precautionary principle did not offer clear guidance on how to choose among important competing priorities or allocate potential risk across more than one desired outcome.

Another alternative reviewed in the environmental impact statement, which reflected the concerns of an alliance of environmental organizations, also included precautionary language. This alternative set strict limits on management options under the assumption that interventions—even those designed to promote forest health—risked causing ecological degradation. In other words, this alternative took the strongly precautionary position of placing the burden of proof of no ecological harm on proponents of any active forest management activity. This strong articulation of the precautionary principle constrains implementation of any form of adaptive management, which assumes that some undesirable effects are likely to occur during the process of learning from experience.

Although neither of these alternatives was selected, their inclusion in the environmental impact statement was clearly a response to precautionary attitudes prevalent among influential stakeholders. Moreover, responding to input from the US Fish and Wildlife Service and environmental groups, the Forest Service incorporated some precautionary guidelines in the policy choice it ultimately selected in the 2001 record of decision (USDA Forest Service 2001b). The selected policy attempted to balance caution and flexibility by emphasizing fire management near human communities and preservation of owl habitat outside these zones. Yet the record of decision also recognized that, because both the policy environment and forest ecosystems are dynamic, some form of adaptive management is necessary (USDA Forest Service 2001b). This awkward compromise soon led to problems in the field as foresters found they could not reconcile the conflicting mandates in practice. Following these complaints, the regional forester commissioned a team of experts to review the selected management plan.

In its 2003 report, the review team confirmed that the effort to balance caution and flexibility was not workable (USDA Forest Service 2003). The team concluded that guidelines included in the record of decision as precautionary measures to protect owls and old-growth habitat actually reduced the likelihood that these goals could be achieved. The team argued, for example, that limits to fuel treatment in areas where owls are present—imposed to minimize damage to owl habitat—had the effect of increasing the risk of catastrophic wildfires in these forest stands, thus heightening rather than reducing threats to the habitat over the long term.

The report also noted a second problem associated with implementation of precautionary approaches: constraints on the ability to take advantage of opportunities to trade off short-term losses for potential long-term gains. If ecological models suggest, for example, that certain forest management and fuel treatment strategies may lead to some losses in owl habitat over the short term but significant potential gains in owl habitat over the longer term, precautionary thinking appeared to preclude such actions.

Suggesting that problems associated with using the precautionary principle as a decision tool are not confined to the Sierra Nevada dilemma, assessments of the Northwest Forest Plan implemented in the Pacific Northwest (Forest Service Region 6) also concluded that application of the precautionary principle had increased the risk of fire in the region, contributed to economic losses in local communities, and dampened innovation in forest management (Thomas 2003; Mealey et al. 2005). Mealey et al. (2005, 199) stated, for example, that in Region 6, "regulating agencies have defaulted to narrow, restrictive actions aimed at the elimination of immediate harm in forest restoration proposals, without simultaneously considering the long-term effects of not doing so. The practical effect has been to allow the declining quality of dry-forest owl habitat to worsen and to continue to expose owls and other resources to unnecessary and preventable risks of uncharacteristic fire."

Conclusions

Precautionary thinking can make important, positive contributions to debates over complex environmental dilemmas. These positive contributions include focusing heightened attention on the long-term effects of policy decisions, the likelihood of unintended consequences, the limitations of cost-benefit and risk assessment methodologies, and the importance of recognizing diverse values among stakeholders and the general public.

Yet problems arise in applying the precautionary principle as a concrete mechanism for selecting among competing policy alternatives in particular cases. These problems occur in part because of mistaken assumptions about the approach. The precautionary principle is a broad statement of desired outcomes, but it is not a concrete policy process or decision support mechanism. To offer an analogy, the adoption of a broad principle of human rights—certainly an important and valuable step—would not help policy makers choose specific, effective strategies for promoting human rights in a particular domestic or international conflict where those rights were threatened. Policy decision frameworks must go beyond statements of principle to include operational processes and analytic tools.

Chapter 5

Adaptive Management

Planners and decision makers often propose an adaptive management approach to deal with scientific uncertainties. As discussed in chapter 4, however, when this approach is combined with the precautionary principle, as was the case in both the Northwest Forest Plan and the Sierra Nevada Forest Plan Amendment, it may lead to unintended negative effects that undermine management outcomes. The experimental nature of adaptive management accepts the possibility of some adverse outcomes. Indeed in adaptive policy making, limited adverse consequences can serve as the basis for learning. Yet this inherent threat of harm can conflict with precautionary approaches, resulting in significant constraints on management options.

As we discuss in detail in this chapter, even without precautionary constraints adaptive management is often inadequate to solve complex environmental problems. Adaptive management that relies primarily on science and ignores value-based components of environmental dilemmas tends to founder on the social and political aspects of the problems. On the other hand, when managers attempt to incorporate stakeholder engagement into adaptive management, they begin to confront the complications and trade-offs associated with public participation that we discuss in more detail in chapter 6.

Practitioners and scholars have explored the concept of collaborative adaptive management—which aims to integrate scientific adaptive management and stakeholder participation—for the past several decades (Holling 1978; Lee 1993; Buck et al. 2001). Variations of the approach have now been implemented in practice to address a range of problems (e.g., Stubs and Lemon 2001; Habron 2003). While adaptive management has made useful contributions in many of these cases, we have not found an example in which adaptive management, whether scientific only, or both scientific and collaborative, has resolved a wicked problem. Adaptive management in one form or another is almost certainly a necessary part of any effort to address complex environmental dilemmas, but its application does not appear sufficient to ensure success. In this chapter, after exploring definitions of adaptive management and considering its strengths and limitations, we review several case studies of environmental conflicts in which adaptive management has been applied.

Definition and History

Adaptive management is a method of addressing complex environmental problems when the outcomes of management interventions cannot be fully understood in advance. Adaptive management emphasizes systematic learning by doing. In this approach, natural resource managers treat policy alternatives and expected outcomes as hypotheses to be tested. Managers implement policies in part as experiments designed to generate critical information that can be used to modify management practices and improve outcomes in the future. Unlike collaborative adaptive management, which we return to later in the chapter, conventional adaptive management focuses primarily on ecological outcomes and does not attempt to account for societal issues related to conflicting public values.

Wilhere (2002) identifies two forms of conventional adaptive management: passive and active. Passive adaptive management involves a process of formulating predictive models, implementing policy decisions based on the models, and revising the models and modifying policies as monitoring data become available. Models are used to predict ecosystem responses and, in theory, management activities can be designed to disturb the ecosystem in ways that provide information to help experts fine-tune the parameters of their models. In passive adaptive management, monitoring and evaluation systems are in place before management begins, but interventions are conducted without controls, replication, or randomization. This has the benefit of making passive adaptive management relatively inexpensive and straightforward to implement. The limitation of this approach is that, without controls, replication, and randomization, it cannot firmly establish cause-and-effect relationships between management actions and changes in ecosystem conditions (Wilhere 2002).

Active adaptive management, in contrast, conducts management actions as deliberate, formal experiments (Wilhere 2002). Alternative policies are viewed as treatments implemented through statistically valid experimental design. Monitoring forms the data collection phase. Active adaptive management can lead to better understanding of how and why natural systems respond to management interventions, and the resulting deeper understanding of environmental responses to a range of treatments can help in designing better policies. The trade-off is that active adaptive management can be considerably more time consuming and expensive to implement.

Incorporating Social Conflicts into Adaptive Management

Conventional forms of adaptive management incorporate scientific methods into the decision process in order to reduce scientific uncertainty. However, difficulties in reaching satisfactory decisions in many complex environmental dilemmas result not only from scientific uncertainty but also from conflicting social values and divergent levels of risk tolerance among stakeholders and the public. In these cases, reducing scientific uncertainty at the margins for experts is unlikely to reduce the larger social conflicts driving the dilemmas. Stakeholders may be as concerned with the political aspects of the management decision-making process as they are with uncertain long-term projected environmental outcomes.

The purpose of collaborative adaptive management is to broaden conventional adaptive management to include social components of the problems. McLain and Lee (1996, 437) argue that "to be effective, new adaptive management efforts will need to incorporate knowledge from multiple sources, make use of multiple systems models, and support new forms of cooperation among stakeholders." Gunderson (1999) finds that, in cases where adaptive management leads to progress, informal, politically independent networks tend to emerge that are effective in part because they operate beyond the narrow world of regulation and implementation. These informal networks are more likely to explore flexible opportunities for resolving resource management issues, devise alternative designs and tests of policy, and develop creative ways to foster social learning. Collaborative adaptive management aims to apply the concepts of adaptive management to the public participation process through the formation of learning networks (Stubbs and Lemon 2001).

Lee (1993) identifies ten conditions that favor collaborative adaptive management:

1. There is a social mandate to take action in the face of uncertainty.
2. Preserving pristine environments is no longer a viable option.
3. Human intervention is unable to produce desired outcomes predictably.
4. There are institutions sufficiently stable to measure long-term outcomes.
5. Researchers are able to formulate hypotheses about the issue.
6. Theory, models, and field methods are available that can allow decision makers to estimate ecosystem-scale behavior.
7. Decision makers are aware that they are experimenting.
8. Organizational culture encourages learning from experience.
9. Resources are sufficient for decision makers to measure ecosystem-scale behavior.
10. Decision makers care about improving outcomes over biological time scales.

While collaborative adaptive management is appropriate in these conditions, the case studies we review below illustrate that it is often difficult to satisfy the institutional conditions in this list.

Adaptive Management in the Context of Complexity and Uncertainty

From the ecological perspective, the term complexity generally implies the existence of multiple levels of interconnected, dynamic relationships among large numbers of interactive agents. In wicked problems, ecological complexity is compounded by social complexity, involving multiple, active, stakeholder groups with diverse values operating in an uncertain and shifting administrative, economic, political, and legal environment.

The presence of layers of ecological and social complexity does not mean that managers should do nothing until the integrated natural and social system becomes stable and well understood and the ramifications of any management action are fully known. Indeed, the presence of complexity means that uncertainty will remain, no matter how much information is collected and analyzed. Bormann and colleagues (1994, 3) argue that "Complexity is [best] confronted by increasing efforts to understand mechanisms that influence change, by reducing expectations that the future can be accurately predicted, and by reducing risk through diversifying." Thus the more uncertainty requires learning and adjustment, the more adaptive management becomes an indispensible part of the management response (Asher 2001).

The presence of uncertainty is central to the concept of adaptive management (Irvine and Kaplan 2001). Unlike nonadaptive management practices, which assume that systems under management are understood and outcomes can be predicted with confidence, adaptive management accepts the reality of incomplete knowledge and focuses on building learning opportunities into policy design and implementation.

As discussed in chapter 4, the precautionary principle has also been offered as a foundation for addressing risk and uncertainty, and there can be pressures for managers to attempt to integrate the two approaches. This can be problematic. As Cooney and Dickson note,

> The relationship between the precautionary principle and adaptive management...is somewhat ambiguous. The relationship, and in particular whether an adaptive management approach should be viewed as a means of implementing the precautionary principle, may depend on the type of potential harm involved (including irreversibility and suddenness of impact) and whether the scenario involves multiple or single sources of risk. If a harm is likely to occur definitively, at a single point in time and irreversibly, an adaptive management approach is likely to be an appropriate means of responding to it in a precautionary manner. (Cooney and Dickson 2005b, 298)

Given that in wicked problems managers by definition are not able to identify which harms are "likely to occur definitively, at a single point in time and irreversibly," we note that neat integration of the precautionary principle and adaptive management is always likely to be elusive in the context of wicked problems.

In practice, the precautionary principle and adaptive management can become opposing policy preferences for factions struggling over resource management, even when members of the competing groups are unfamiliar with the terms as defined by experts and scholars. Those opposed to economic development of natural resources, for example, may argue for a precautionary approach, especially a *no action* alternative letting nature take its course, while those favoring development may argue that the environmental risks of economic development can be managed through improved technology or modifications of policy over time (Tucker and Treweek 2005). In wicked problems, these views resist reconciliation.

Examples of the Use of Adaptive Management

To evaluate the successes and pitfalls of adaptive management, we reviewed twelve cases described in the literature. The cases involved the actual, or attempted, application of adaptive management to the following issues: air quality in Great Britain (Stubbs and Lemon 2001); salmon in the Columbia River basin (Smith et al. 1998); watersheds in southwestern Oregon (Habron 2003); waterfowl harvests in North America (Johnson and Williams 1999); water projects in Australia (Gilmour, Walkerden, and Scandol 1999); water allocation in the Florida Everglades (Walters, Gunderson, and Holling 1992); forests and biodiversity in the Pacific Northwest (Stankey et al. 2003); forests and biodiversity in the Sierra Nevada (Thomas 2003); riparian and coastal ecosystems in the United States (Walters 1997); water use at Fort Huachuca in California (Gen 2001); flood cycles at the Glen Canyon Dam in Arizona (Meretsky, Wegner, and Stevens 2000); and salmon fisheries in British Columbia (Pinkerton 1999).

In all of these cases, managers attempted to incorporate some level of stakeholder collaboration or broader public participation into adaptive management. In combination, outcomes in these cases suggest lessons regarding the use of adaptive management to address complex environmental dilemmas.

We review the first eight of these cases in some detail. The studies of the Everglades and the Sierra Nevada are directly relevant to two of the four wicked problems that we refer to throughout the book. In this chapter's concluding section, we summarize lessons regarding the application of adaptive management from all twelve cases.

Air Quality Standards in Great Britain

In the United Kingdom in the mid-1990s, a national effort to improve air quality began with the assumption that local governments were the most appropriate level of authority for implementation. This reflected a movement toward devolution across many areas of government action. Yet because air-quality problems transcend jurisdictional boundaries, an approach based on local authority would require unprecedented cooperation and collaboration within and across jurisdictions. Moreover, the proposed national approach took the form of an unfunded mandate. That is, the plans did not include additional financial support from the national government, making local officials apprehensive about taking on the new responsibilities.

In a study of local government responses to the proposed program, Stubbs and Lemon (2001) examined the particular case of Bedfordshire County. In collaboration with local officials, the researchers first mapped the network of relevant agency stakeholders and then conducted workshops to develop and strengthen an adaptive process that members of the network could use in response to air-quality management issues. The workshops revealed that participants brought with them divergent and occasionally conflicting interpretations of the problem and the proposed process to address it. Often these differences were correlated with the participants' organizational affiliations. The workshops also revealed that a lack of prior cooperation among the participants, and their respective agencies, was a potential obstacle to effective collaborative adaptive management of the complex issue before them.

Nevertheless, the researchers reported some success in overcoming these obstacles through the use of facilitated stakeholder deliberation and "creative dialogue" in the workshops (Gordon 1961; Senge 1990). Stubbs and Lemon (2001) concluded that the process resulted in the development of a virtual overarching organization of stakeholder representatives that transcended the divisions linked to participants' initial strong self-identification with their actual organizations. This transorganizational network demonstrated a strengthened ability to engage in "joined-up thinking," which, in turn, led to management processes that were more responsive and adaptive (Stubbs and Lemon 2001, 329).

Based on their work, the researchers asserted that the common approach of assigning responsibility for management of a given complex problem to a particular agency constrains opportunities for collaborative adaptive management. The authors also reported that the few individual government employees participating in their workshops whose formal job descriptions included interagency communication or cooperation had a disproportionate positive impact in building the adaptive learning network that emerged. The researchers emphasized that these organizational "boundary diplomats" are often best able to perform a "difficult but essential role in facilitating a more holistic and less parochial approach" (Stubbs and Lemon 2001, 324).

Salmon in the Columbia River Basin

Beginning in the 1980s, the Northwest Power Planning Council had substantial responsibility for addressing the problem of salmon management in the Columbia River basin in the northwestern United States. The council explicitly adopted adaptive management in 1984 as a way to organize its activities to protect migratory salmon populations. In combination, the complex life cycle of the fish species, the uncertainty of the scientific data, the divergent views of the value of the resource to many different stakeholders, and the multiple stressors to the health of the species linked to human activities created a situation where conflicts abounded and blame shifting was common.

Kai Lee, a member of the council, initially proposed the application of a collaborative form of adaptive management for addressing threats to the salmon populations. In introducing the approach, Lee (1993) used the metaphors of compass and gyroscope to help explain how science and democracy could be integrated in adaptive, deliberative decision making. According to these metaphors, when policy makers are navigating in unknown waters, scientific findings can be a compass pointing the way, while public participation through democratic processes can be a gyroscope maintaining stability. Unfortunately, as Smith and others (1998, 671) note in a retrospective assessment of the application of adaptive management in the case of salmon in the Columbia River basin, "the scientific compass often points in more than one true direction, and the public gyroscope tilts precariously with political turbulence."

In their assessment of salmon management decision making, Smith and colleagues (1998) examined the role of scientists and scientific studies and also reviewed stakeholder and public attitudes. They found multiple drivers of misunderstanding, miscommunication, and conflict. They note that scientists involved in the salmon issue often tried to be both compass and gyroscope, thereby confusing their roles and the public's perceptions of their findings. Smith and colleagues (1998) point out that science is better at showing which policy directions are wrong than in determining with certainty which direction is right. Thus science is not well suited to "proving that new methods will work" (Smith et al. 1998, 676). Members of the public rarely fully understand this distinction.

The researchers also found that this misunderstanding of science, and of the meaning of scientific disagreement, ultimately fueled public distrust of experts. Because the results of scientific inquiry are often not communicated effectively to the general public, the public's understanding of the available knowledge and its limitations is incomplete. Consequently, various stakeholder groups were able to use scientific findings that appeared contradictory to support their divergent positions and paradigms (Smith et al. 1998).

Not surprisingly, given the scientific uncertainty, Smith and colleagues (1998), in reviewing surveys of stakeholder attitudes toward the salmon problem, found that the public, users, managers, and scientists did not agree on the causes of salmon decline. But they also believed that in practice they could not influence decisions. Smith and colleagues (1998) concluded that public involvement in the Pacific Northwest salmon issue was inadequate. They wrote that "Too often, public participation merely fulfills a legal requirement rather than helping to improve decision making" (Smith et al. 1998, 679).

The Columbia River basin adaptive management effort had some initial success as a result of strong leadership by the Northwest Power Planning Council and available funding for experimentation through an annual rate-payer assessment on power usage (Lee 1989; McLain and Lee 1996). The process ultimately struggled, however, as stakeholders groups were unable to work together and litigation under the Endangered Species Act ensued (Volkman and McConnaha 1993; Lee 2001).

Watersheds in Southwestern Oregon

Oregon has encouraged the establishment of local watershed management councils to help improve water quality in streams and rivers and support salmon conservation. The councils facilitate landowner participation to achieve these goals. State guidelines for the councils mandate broad community participation and official local government recognition. The decline of salmon populations in the Pacific Northwest was the primary incentive for establishing local watershed councils (Habron 2003). Oregon's policy favors the councils as a buffer between local landowners and state and federal regulatory efforts, which typically generate significant resistance in areas where both distrust of government and belief in private property rights are strong.

The state offers technical and financial support to the councils. State biologists are available to advise the councils on conservation priorities and methods, and the Oregon Water Enhancement Board offers a competitive grants program to underwrite council projects. Douglas County, Oregon, officially recognized the Umpqua Basin Watershed Council in March 1997.

As required by state guidelines, the council's sixteen-member management committee represents a range of stakeholders. These include landowners who, in this area, are primarily involved in agriculture and grazing, and representatives of fishing and conservation groups, public utilities, and the county government. The council meets once a month and makes decisions by consensus.

According to Habron (2003), the Umpqua Basin Watershed Council identified several key objectives as it started its work: to reduce bureaucracy; to foster productive stakeholder discussion and understanding; and to provide financial, technical, and coordination support to watershed management activities. In bringing science to bear, the council agreed to consider the so-called Bradbury process, named after the steps recommended by biologists in a handbook commissioned by Oregon state senator Bill Bradbury (Bradbury et al. 1995). The Bradbury process outlines a "scientifically-based framework for prioritizing native salmon and watershed protection and restoration activities...[and] provides sound scientific grounding to meet policy needs" (Bradbury et al. 1995, 3).

Habron (2003) examined the operation of the Umpqua Basin Watershed Council as a case study of a community-based approach that aimed to integrate science and diverse public values through collaborative adaptive management. His findings revealed many challenges to success (Habron 2003). The primary obstacle as reported in the case study grew from council members' entrenched beliefs and resistance to compromise. In principle, a diverse community-based council applying collaborative adaptive management should be able to enhance mutual understanding among competing points of view and thereby reduce conflict and encourage cooperation. In practice, however, as Habron (2003) reports, the council divided along a fault line, with private property interests on one side and watershed and salmon conservation interests on the other. This sharp division made it impossible for the council as a whole to view watershed management projects as opportunities for learning.

The author gives the example of an impasse over the best way to keep livestock from damaging fragile riparian zones. The council members with interests in conservation and the health of fish populations proposed a requirement that landowners control their livestock with fencing. The landowners instead favored a voluntary approach encouraging the installation of water sources for livestock away from streams and rivers. As Habron describes it, the contentious and long-running debate over this issue focused on the abstract pros and cons of the ideology of property rights. The council could never agree to move forward to actual policy experiments under an adaptive management approach, which might have provided new knowledge on the strengths and weaknesses of the two proposed alternatives. Habron also notes that the council's policy of making decisions through consensus strongly favored the status quo. Any disagreement among council members had the effect of blocking action.

The council also was divided on the role of science in determining how and where restoration efforts should be focused. Habron reports that biologists and other experts were hesitant to work with the council because the polarized politics made it unlikely that their participation would be useful. Consequently the council was unable to implement the science-based Bradbury process that it had initially agreed to consider. Thus during the period of study the council was unable to break through the obstacles to collaborative adaptive management. Habron (2003, 39) concluded that "an explicit effort will be required to infuse adaptive management concepts into the councils involved in watershed planning. Without significant council use and participation, adaptive watershed management in watersheds dominated by private lands holds little promise and only continues the current trend of incomplete adaptive watershed management in the Pacific Northwest."

Waterfowl Harvests in North America

Long-standing problems in waterfowl harvest management led the US Fish and Wildlife Service to adopt an adaptive management approach in 1995 (Johnson and Williams 1999). These problems included unclear goals, lack of objectivity in decision making, underestimation of the significance of uncertainty in management outcomes, and inadequate monitoring and policy adaptation.

In response, scientists at the agencies responsible for waterfowl management created a technically advanced program to estimate optimal harvest regulations using both passive and active adaptive management protocols. Johnson and Williams (1999) and other experts developed recursive algorithms to support Markov decision processes. Markov processes provide mathematically based decision criteria for policy makers in situations where outcomes are a function of both management choices and random variability in the system. In developing these protocols, the scientists made substantial progress in incorporating into their models four key sources of uncertainty: (1) uncontrollable natural environmental variation, (2) lack of precise agreement between the intended and actual harvest limits resulting from management actions, (3) lack of complete understanding of dynamic biological and ecological processes, and (4) lack of precision and comprehensiveness in monitoring and data collection.

Johnson and Williams (1999, 2) remarked that in pursuing this work they "benefit from an institutional commitment to adaptive management that so often has eluded other natural resource managers." The more common circumstance in environmental dilemmas is that adaptive management is not implemented, despite lip service, because of the short-term risks of adverse outcomes from policy experimentation. Agency leaders subject to political pressures are often reluctant to allow the use of management choices explicitly for learning, rather than for optimizing outcomes given current knowledge, even though staff scientists frequently argue that outcomes will improve over time if systematic and rigorous policy experimentation is permitted.

Despite their progress in developing and implementing technical adaptive management protocols, Johnson and Williams (1999) recognize the limitations of their approach when overall management goals are ambiguous. They write that the type of adaptive management they describe "is not a process for coping with disagreement over management goals and objectives" (Johnson and Williams 1999, 6). They observe, for example, that waterfowl harvest managers typically assume that hunter satisfaction is closely linked to the number of birds taken and thus set regulations to promote hunting success. Yet recent work to assess participant preferences reveals that the key sources of satisfaction for waterfowl hunters tend to be the social and aesthetic aspects of the activity, rather than simply the number of birds taken. The authors also point out that the ongoing movement toward expanded aboriginal rights in Canada, a key part of migratory waterfowl flyways in North America, is bringing new stakeholders into the decision-making process, including aboriginal people who wish to hunt for subsistence and for cultural reasons and the provincial officials who will have to adapt their policies as a result.

As Johnson and Williams conclude:

> Ultimately, the success of adaptive harvest management depends on a general agreement among stakeholders about how to value harvest benefits and how those benefits should be shared....It is these unresolved value judgments, and the lack of effective institutional structures for organizing debate, that present the greatest threat to adaptive harvest management as a viable means for coping with management uncertainty. (Johnson and Williams 1999, 8)

Water Projects in Australia

Gilmour and colleagues (1999) reviewed three cases of the application of adaptive management to address water use and water quality issues in areas with growing populations around Sydney in New South Wales, Australia. In their review, the authors looked particularly at adaptive management's role in educating participants, promoting negotiation, and providing a framework for policy formulation and evaluation. Each of the three cases involved representatives of the local government bodies with management responsibility and affected stakeholders.

In the first case, participants in a series of workshops over a two-week period examined the negative environmental effects of urban development in an area southwest of Sydney and considered a range of policy responses. The participants used a simulation model of land-use and water-management alternatives to assess water management, land management, and socioeconomic issues and to capture uncertainties linked to population growth and management effectiveness. The primary sources of degraded water quality were sewage and urban runoff. The participants concluded in their review that, even after accounting for uncertainties, existing land-use patterns and proposed future developments would necessarily result in failure to meet water quality standards.

The workshops were successful in developing near consensus on desired policy outcomes and in promoting collaboration between the regional water board and community stakeholders. Based on a report of the results of the workshops, the staff of the board in collaboration with workshop participants developed procedures for implementing adaptive management of water quality. The regional manager of the water board became the project's *institutional champion,* supporting the outcomes. This progress was abruptly aborted, however, as a result of a shift in the external political environment. New changes to the water board's charter required the institution to emphasize corporate-style provision of services over resource management. As a result, the adaptive management approach to water quality was abandoned.

The second project focused on management of several lakes in a coastal zone affected by urban expansion to the north of Sydney. The lakes had a history of eutrophication linked to nutrient flows from sewage and agricultural runoff. Prior policies had included the installation of filtering wetlands and the dredging of openings to the sea to promote tidal flushing. The project described by Gilmour and colleagues (1999) involved a series of workshops and community meetings to address these issues. Workshop participants included community members, elected officials and staff from the local council, representatives of relevant government agencies, and technical experts.

Participants agreed that eutrophication of the lakes continued to be the primary water quality concern. In response they focused on improving their understanding of the sources and flows of nutrients and sediments, and on assessing the management options for dealing with these problems. Participants also worked to characterize uncertainties, particularly related to the ecological effects of the entrance channel to the ocean, the effectiveness of storm water-management installations, and the dynamics of nutrient and sediment flows. The workshops led to the development of a decision support model for evaluating water quality management strategies. In identifying policy strategies, participants emphasized the importance of controlling nonpoint sources of pollution and improving nutrient flushing.

By the end of the workshops and community meetings, participants had agreed on a general management strategy but had not yet clarified concrete steps or begun implementation. Once again, unexpected external influences intervened. In this case, the agency's institutional champion transferred away. A member of the consulting team managing the workshops took on the role to maintain internal support as the project moved to implementation. In their review, Gilmour and colleagues (1999) reported that cooperation among community members and the local county council was effective. On the other hand, they noted that little learning took place and the effort was handicapped by less than full commitment from government agencies.

The third case involved the catchment area for Sydney's water supply reservoirs. The effort in this area was undertaken in response to pressures for increased access to the lands around the reservoirs from public groups, including conservationists, hikers, four-wheel drive clubs, and ideological advocates for open access to public lands. The water board was concerned about the effects of greater access on water quality. Complicating matters, the various groups seeking increased access had a contentious history of mutual antagonism and distrust.

In a series of workshops, representatives of the groups lobbying for access, along with managers and scientists, met to assess the issues, consider management alternatives, and offer recommendations. Given the hostilities among representatives of the groups, the workshop facilitators urged participants to separate themselves from their affiliations and to work initially toward the goal of consensus on the importance of maintaining water quality while permitting appropriate levels of access. The scientists made presentations to clarify technical issues.

On the divisive question of the impacts of various recreational activities on water quality, the participants learned that causal relationships were actually poorly understood and that scientists did not have adequate data on reliable indicators. This mutual recognition of uncertainty led the participants to agree on the need for collecting and analyzing new data on both the impacts of recreational activities and the costs and benefits to stakeholder groups of access decisions (Gilmour, Walkerden, and Scandol 1999). The group further agreed to recommend that the existing access limits should remain in place until the studies were completed.

In their review, Gilmour and colleagues (1999) argue that in all three cases collaborative adaptive management strategies showed promise. In particular, participants learned about the challenges of decision making under uncertainty, expanded their appreciation of opposing positions, and contributed to productive negotiation. Recurring obstacles to progress also emerged. These included a diminution of support from local and national government agencies once the level of political controversy died down, general problems in sustaining long-term follow-through from all participants, and the disruptive effects of unexpected external changes in personnel or political support (Gilmour, Walkerden, and Scandol 1999).

The Florida Everglades

In the late 1980s, professional and technical experts began to express growing concern about a number of significant chronic problems related to natural resource management in the Florida Everglades. Over a period of two and a half years, a group of fifty experts met in a series of workshops to share information about the ecosystem in an effort to strengthen understanding and thereby support improved management choices. During this informal, collaborative effort, the scientists worked to develop computer models to simulate the ecosystem's spatial and temporal dynamics (Walters, Gunderson, and Holling 1992). This was a challenging task. Models of some system components, particularly the hydrological models, gained credibility as they were able to match existing historical data. Others, particularly those attempting to capture trends in biodiversity and interactions among species, were less convincing.

After some time, while acknowledging that they still lacked a complete understanding of ecosystem dynamics, the participants concluded that they had sufficient information to offer management recommendations. Six of the participants developed a new proposal for ecosystem restoration. With the proposal ready, they worked to expand participation beyond the group of experts involved in the workshops. First, they prepared materials to help communicate to nonexperts the technical understanding they had achieved. These materials included a computer animation of the hydrology of the Everglades and a set of brief documents discussing in lay terms what was known and not known about key issues. Second, they brought this material to policy makers and other stakeholders. They held individual meetings with board members of the South Florida Water Management District and a workshop with representatives from the public and private sectors and nongovernmental organizations.

In some ways, however, this process, driven by skilled, experienced, and well-meaning technical experts, led to the problem described in the case of the salmon in the Pacific Northwest. That is, scientists found themselves trying to be both compass and gyroscope. They attempted, largely without success, to incorporate collaborative adaptive management into their activities. Many adaptive management policies were recommended, but none were implemented (Gunderson et al. 1995; Gunderson 1999). Three factors contributed to this outcome: (1) limited flexibility in the social system; (2) little or no resilience in key components of the ecological system; and (3) technical challenges with designing experiments.

Proposals for adaptive management were unable to overcome the inflexibility of entrenched relationships among stakeholders and management agencies and the narrowly interpreted legal mandates that constrained agency actions. Stakeholders who benefit from the status quo could block alternative management regimes through real and threatened lawsuits. Decision makers typically did not have legal leeway to experiment with resource management practices. Administrators cannot apply adaptive management when the associated risks of failure are socially and legally unacceptable. Gunderson writes that,

> until management institutions are capable and willing to embrace uncertainty and to systematically learn from their actions, adaptive management will not continue in its original context, but will be redefined in a weak context of "flexibility in decision making." In cases of successful adaptive assessment and management, an informal network seems always to emerge. That network of participants places emphasis on political independence, out of the fray of regulation and implementation, places where formal networks and many planning processes fail. The informal, out of the fray, shadow groups seem to be where new ideas arise and flourish. It is these "skunkworks" that explore flexible

opportunities for resolving resource issues, devise alternative designs and tests of policy, and create ways to foster social learning. (Gunderson 1999, 5)

Unfortunately, as the Everglades case suggests, the flexible and adaptive proposals emerging from informal networks often struggle to survive the inflexible social and legal limits that constrain effective implementation of innovative management approaches. Barriers to adaptive management arise when there is limited resilience in natural systems and entrenched inflexible power relationships among stakeholders. The lack of institutional flexibility appears to contribute to a wide range of poor management performances.

The Northwest Forest Plan

The record of decision for the Northwest Forest Plan (Interagency SEIS Team 1994) provided guidance for managing the twenty-four million acres of federal lands administered by the Forest Service, Bureau of Land Management, and National Park Service within the range of the northern spotted owl. The selected management alternative was a precautionary choice that emphasized avoidance of ecological risks to spotted owls, other species, and their habitats on federal lands in the region. The decision created a network of reserves that expanded the coverage of existing protected zones, such as national parks and wilderness areas, to encompass more than three-quarters of the land addressed by the Northwest Forest Plan. In these areas, management interventions and development activities were sharply limited. In addition, a set of comprehensive and prescriptive standards and guidelines promulgated with the plan further constrained management and development activities, both within and outside the reserves.

Despite this emphasis on precautionary approaches in managing the great majority of the land, the Forest Ecosystem Management Assessment Team, created by President Clinton to address concerns about the decline of old-growth forests and associated wildlife, recommended some application of adaptive management to address uncertainties arising from ecological and political complexity. The Northwest Forest Plan implemented this recommendation by establishing several Adaptive Management Areas. These areas, encompassing about 6 percent of the federal lands within the planning area, were set aside for experimentation with more innovative and flexible methods for achieving social and ecological goals. The idea, following the precepts of adaptive management, was to allow the agencies to learn from policy experimentation to provide a basis for strengthening future management efforts (Interagency SEIS Team 1994).

Stankey and colleagues (2003) evaluated the adaptive management portion of the Northwest Forest Plan. They interviewed many of the key agency personnel involved in implementing the Northwest Forest Plan and the Adaptive Management Areas. In their research, they found generally that the effort to apply adaptive management as part of the Northwest Forest Plan had failed. Based on their interviews, they identified several key barriers to success.

- *Leadership and definitional barriers*: Area coordinators and line officers given the task of implementing adaptive management did not have clear guidance and support from the agency leadership. Administrators at the Forest Service regional office explicitly decided not to provide directions for implementing adaptive management, intending to avoid imposing top-down rules. But because they did not communicate this decision to the field staff, the regional office's silence had the unintended effect of indicating a lack of support for adaptive management. Further, those charged with implementing adaptive management had differing conceptions and expectations regarding its definition, purpose, and objectives.
- *Institutional barriers*: Federal land management agencies are generally risk averse. A policy climate exists in which land managers are expected to avoid any management actions with the potential to cause harm to threatened and endangered species or their habitat. This precautionary policy climate is clearly contradictory to the concept of adaptive management. Consequently, without clearly defined and articulated incentives, the adaptive management area coordinators and line officers had no expectations that they would be rewarded for the risk-taking inherent in adaptive management. Further, as budgets declined, funds were diverted away from the adaptive management areas to support other operations associated with the Northwest Forest Plan. The default tendency for agencies to adhere to prescriptive approaches and standardized rules constrained innovation, even when specific directions authorizing experimentation within the adaptive management areas was provided.
- *Statutory and regulatory barriers*: Laws governing natural resource management and biodiversity protection, such as the Endangered Species Act, generally allow little flexibility, and the responsible

regulatory agencies, such as the Fish and Wildlife Service, enforce these laws strictly. The laws do not contemplate application of adaptive management, a concept that only emerged relatively recently. Consequently, the regulators and the courts are unlikely to approve adaptive management actions because of the possible risk of experimental policies to listed species. This holds true even when the status quo also holds risks for species and habitat and even if, as in the case of adaptive management, the object is to improve long-term outcomes through learning. In effect, adaptive management cannot be implemented unless risk can be eliminated. As Stankey and colleagues (2003, 44) write, "The resulting catch-22 situation, in which experimentation can be undertaken only if there is a guarantee of no adverse consequences, establishes a difficult, if not impossible, decision criterion to satisfy."

- *Barriers to learning*: Despite the central importance of learning by doing in adaptive management, the agencies provided no clear definition of learning or any explanation of how institutional learning would take place as a result of activities in the adaptive management areas. Moreover, the area coordinators received no special training or support staff, and their regular responsibilities severely limited their opportunities and incentives to develop and apply formal adaptive management experiments. After some time, any management activity that occurred in the adaptive management areas was deemed to be adaptive management, whether or not the activity had any connection to experimentation and institutional learning.

In a similar study with similar findings, Thomas (2003) examined performance in the two adaptive management areas in northern California. This study concluded that management practices in the adaptive management areas were no different from those applied elsewhere on Forest Service lands. Thomas (2003) writes, "Testing of S&Gs [standards and guidelines] has become less acceptable due to both 'internal and external resistance.' There has been a constant degradation of the flexibility that was to be a part of AMA [adaptive management area] activities due to internally created barriers."

Lee (2001), in an appraisal of adaptive management as applied by the Forest Service, notes that the agency uses the following definition of adaptive management in the Northwest Forest Plan: A method "for land managers, researchers and communities [to] work together to explore new methods of doing business" (USDA Forest Service 1998). Lee (2001, 12) concludes that "the Forest Service definition of adaptive management does not emphasize experimentation but rather rational planning coupled with trial and error learning. Here 'adaptive' management has become a buzzword, a fashionable label that means less than it seems to promise." Mealey and colleagues (2005) further argue that implicit acceptance of the precautionary principle in the Northwest Forest Plan interfered with the application of adaptive management, with adverse consequences for biodiversity and habitat.

The Sierra Nevada National Forests

The Sierra Nevada Forest Plan Amendment process was closely tied to the controversies that led to the Northwest Forest Plan, and the outcomes for adaptive management were similar. In the 2001 record of decision for the Sierra Nevada Forest Plan Amendment process, the regional forester attempted to balance the conflicting strategies of caution and flexibility. The selected policy alternative, for example, took a precautionary approach emphasizing fire management within the wildland-urban interface zones near developed areas (Stewart et al. 2007) and a precautionary approach emphasizing preservation of California spotted owl habitat outside these zones. Yet the record of decision also recognized that because neither management strategies nor forest ecosystems remain static, some form of adaptive management was essential (USDA Forest Service 2001b).

As in the Northwest Forest Plan, however, the precautionary thinking inherent in the development of the standards and guidelines accompanying the record of decision prevented effective application of adaptive management. A team appointed by the regional forester to review implementation of the original decision concluded that (1) the prescriptive standards added to protect California spotted owls and old-growth habitat actually reduced the likelihood that the goals stated in the record of decision could be achieved, and (2) precautionary constraints prevented the opportunity to take advantage of possible trade-offs of modest short-term losses for more significant long-term gains (USDA Forest Service 2003). As mentioned earlier, Thomas (2003) found similar conflicts between the precautionary principle and adaptive management in his review of adaptive management areas established in California under the Northwest Forest Plan. As we discuss in chapters 3 and 8, the Forest Service ultimately had to revisit the 2001 decision in the Sierra Nevada case in part as a consequence of these built-in inconsistencies and contradictions.

Other researchers have drawn similar conclusions regarding the challenges to effective implementation of adaptive management given the institutional constraints of the Forest Service. Butler and Koontz (2005) surveyed district rangers, forest supervisors, and regional foresters regarding their perceptions of the agency's implementation of ecosystem management, another promising approach to natural resource management. The authors write,

> Agency managers perceive that the greatest attainment of such objectives is related to collaborative stewardship and integration of scientific information, areas in which the organization has considerable prior experience. The objectives perceived to be least attained are adaptive management and integration of social and economic information, areas requiring substantial new resources and a knowledge base not traditionally emphasized by natural resource managers. (Butler and Koontz 2005, 138)

Conclusions

The logic of applying adaptive management to address uncertainty is compelling. The process of policy experimentation can lead to new knowledge to help improve future management decisions. And while adaptive management does bring some risk of adverse outcomes as managers learn from experience, nonadaptive management under conditions of uncertainty also poses significant risks. Indeed, under conditions of uncertainty, risks from adaptive management may be lower than risks from conventional approaches, because in adaptive management the risks are explicitly recognized and can be controlled by limiting the scale of the experiments and applying careful monitoring of outcomes (Moir and Block 2001).

Yet the case studies reviewed in this chapter indicate that it is much more difficult than anticipated to implement adaptive management in practice. First there are significant structural constraints. Environmental laws, regulations, court rulings, and political pressures all conspire to foster risk aversion in natural resource management agencies. Even if organizational leaders favor experimentation and learning, they are often not legally permitted to try innovative and flexible—but potentially risky—options. Conversely, even if legally permitted to apply adaptive management, the agencies typically lack the experience, know-how, organizational culture, and financial resources to do so effectively.

These conditions create what Stankey and colleagues (2003, 44) call the catch-22 situations associated with implementation of adaptive management in the current legal, regulatory, and political climate. As Gunderson (1999) observes, creative ideas for adaptive management often come from relatively small groups of scientists or other practitioners working together informally, but translating their ideas into practice, given the structural constraints described earlier, is a daunting challenge. Allan and Curtis (2005) find that both active and passive adaptive management are constrained by entrenched social norms and institutional frameworks that create "barriers to change" and "barriers to adoption."

These structural barriers apply whether the adaptive management plans are primarily scientific and technical or whether they also incorporate stakeholder collaboration and deliberation. But the case studies reviewed here reveal that collaborative adaptive management faces a second set of formidable obstacles. While several of the studies document some progress in deliberation and negotiation as stakeholders with divergent views worked together to deal with environmental controversies, none reports ultimate success in untangling the deep-seated disagreements and achieving implementable and sustainable policy solutions. Instead the researchers describe the emergence of a variety of problems. These include the ability of individuals or small groups to stall action even when the majority of participants are in agreement, the reluctance of technical experts to engage in contentious political discourse, the lack of experience in collaboration by agencies and stakeholders, the lack of public understanding of scientific methods and findings, the detachment of management agencies, the turnover of key agency personnel and stakeholder participants, and the threat of unpredictable disruptions from changes in external political, budgetary, or other conditions.

These results suggest that adaptive management may have more promise for addressing scientific uncertainty than for addressing divergence in public values (Johnson and Williams 1999; Allan and Curtis 2005). Other scholars (Pahl-Wostl et al. 2007; van der Brugge and van Raak 2007) suggest that transition policies to lay a firmer foundation for collaborative adaptive management may be necessary before the approach can succeed.

Despite the generally pessimistic views of the approach as currently practiced, however, several studies also offer constructive lessons that may be valuable for practitioners and participants in collaborative adaptive management going forward. In a review of citizen involvement in national forest planning, Shindler and Cheek (1999) suggest that,

> Citizen-agency interactions are more effective when (1) they are open and inclusive, (2) they build on skilled leadership and interactive forums, (3) they include innovative and flexible methods, (4) involvement is early and continuous, (5) efforts result in action, and

(6) they seek to build trust among participants. Particular attention to the situational context of actions and decisions helps to determine the relevance of adaptive management for individuals in these settings. (Shindler and Cheek 1999, 1)

Bormann and colleagues (1994) emphasize the importance of engaging a full range of public stakeholders, drawing out their preferences, and encouraging them to think creatively. These authors write that,

> Because societal values and ecological capacity must be integrated to achieve sustainability, iterative interaction is required. Defining what is ecologically possible is not efficiently pursued without first knowing what people want; reconciling what is desired with what can be sustained and finding creative solutions require understanding what is and is not thought to be possible. Thinking about new possibilities does not come easily, and ideas from a broad cross-section of society are needed (Bormann et al. 1994, 8)

Olsson and colleagues (2004) urge public managers to pay as much attention to the social and policy environment as they do to the natural environment for which they are professionally responsible. These authors recommend that,

> The institutional and organizational landscapes should be approached as carefully as the ecological in order to clarify features that contribute to the resilience of social-ecological systems. These include the following: vision, leadership, and trust; enabling legislation that creates social space for ecosystem management; funds for responding to environmental change and for remedial action; capacity for monitoring and responding to environmental feedback; information flow through social networks; the combination of various sources of information and knowledge; and sensemaking and arenas of collaborative learning for ecosystem management. (Olsson et al. 2004, 75)

To these recommendations, we add several others drawn from our review of the literature and our detailed study of the Sierra Nevada case.

- There need to be profound cultural shifts in the agencies and the public leading to tolerance for reasonable risk-taking in environmental management.

- The responsible agency needs to be committed to stakeholder engagement and adaptive management, and there should be sufficient funding to support the agency's long-term commitment.
- All stakeholders, including decision makers in the responsible agency's chain of command, and representatives of other relevant federal, state, and local agencies, need to be involved in decision making.
- The entire decision-making process needs to be open, transparent, and iterative, incorporating and responding to stakeholder values.
- There needs to be reliable funding for policy experimentation and monitoring over the long term (more than ten to twenty years). In addition, the monitoring process should be fully transparent to allay public distrust of agency actions.
- Adaptive management experiments should start at small scales to keep initial costs and ecological risks low, build trust, strengthen learning networks, and provide short-term results critical to maintaining stakeholder interest.

A crucial problem related to this last point is that, unless existing constraints can be overcome, scaling up adaptive management to the ecosystem level will be too costly and too time consuming to be practical. More than fifteen years have passed since the start of the Sierra Nevada Forest Plan Amendment process, and the wicked problem is still generating challenges from multiple stakeholders. Under an institutionalized adaptive management process, decisions could be designed to be revisited every five years as a matter of policy, but at the ecosystem scale the potential environmental impacts would be so large that under existing regulations a comprehensive environmental impact statement would be required each time. The current timeline for completing an environmental impact statement is two or more years, so without regulatory changes, the planning process would in effect become continuous, with implementation always delayed.

In conclusion, while in principle adaptive management would likely be an improvement over current institutionalized practices for environmental management in the face of uncertainty, research on outcomes in the field lead to a sobering view of the obstacles to effective implementation of adaptive management in practice. As we have discussed, many of these challenges are linked to problems of public participation and stakeholder engagement. We consider these issues next in chapter 6.

Chapter 6

Participatory Processes

The goal of this chapter is to both review the best practices of participatory processes and outline an approach that provides sufficient and appropriate participation within the context of wicked environmental problems. It is now part of the received wisdom that public participation is essential in managing complex environmental problems. Such participation is both intended to elicit (at least implicitly) broadly held public values relevant to the management decision at hand, and to incorporate those values into the final decision. However, because typical participatory processes generally fall victim to shortcomings that limit their utility in dealing with wicked problems, decision makers are often frustrated with, and question, the ultimate benefits of public participation.

One shortcoming is that because public participation is generally limited in scale, documented successes most commonly occur in a single community (Shindler and Cheek 1999). There are few if any successful participatory processes at the scale of the Sierra Nevada, a state the size of Florida, a nation, or across national boundaries.

A second limitation is that decision-making authorities fail to maintain participatory processes over time. Given the cost in both time and money of significant participatory processes, they are most commonly used to obtain public input at a given point in time. If the decision process demands sustained public engagement, participation generally is limited to a small set of participants who, managers hope, represent the diversity inherent in the community. To successfully manage wicked problems, decision makers must have both an understanding of broadly held public values and priorities, and a recognition that those value positions will change over time and in response to new scientific knowledge. Thus the participatory approach must enable decision makers to elicit and incorporate public values relatively quickly and at a cost that can be institutionalized and sustained into the future. Stakeholders must assess public values within weeks or months, not years. And the approach must make it feasible to reassess and identify changes in those broadly held values on a regular basis.

Participation in Environmental Management

An essential component in making environmental decisions, especially in the case of wicked problems, is having broadly based social groups involved, beyond specialized communities of experts and political leaders. National and subnational governments routinely require public input in the process of developing policy and managing the environment. The US National Environmental Policy Act and the US Federal Advisory Committee Act both mandate public participation. To cite just one other example, in the cover letter to the report detailing its "model plan for public participation," the US Environmental Protection Agency's Public Participation and Accountability Workgroup states, "The National Environmental Justice Advisory Council considers public participation crucial in ensuring that decisions affecting human health and the environment embrace environmental justice" (NEJAC 2000). Such participation has been a major objective of European and American environmental policy (Renn 2006). Participation in environmental decision making is now increasingly regarded as a democratic right (Reed 2008), and requirements for public participation are *enshrined* in international legislation (Petts and Brooks 2006; Webler and Tuler 2006).

Unlike the precautionary principle or adaptive management, calls for greater participation in environmental decision making are about the *process* of managing the environment, not the *principles* that should guide management decisions. In fact, one of the key beliefs motivating many efforts to increase participation is the view that different stakeholders are motivated by very different principles, and unless a broad base of participants gets involved, key value perspectives may be critically underrepresented in the decision process (Reed et al. 2009).

Public participation lies along a continuum of democratic practice, ranging from *pure representative democracy* to *pure democracy*. Only rarely has either extreme proven workable for states or nation-states. Consequently, in actual practice the level and nature of public participation fall somewhere along this continuum. It is important to note that (1) the amount of participation acceptable in a given context is not static, and (2) over the past several decades the level of participation expected by the public in many societies has substantially increased.

One of the most prominent factors contributing to the growth in participatory decision making has been the realization that many policy decisions are not merely scientific and technical in nature, but also political and social, thus requiring analysts and decision makers to assess interests and values not easily reflected or incorporated in traditional analytic techniques. This observation is particularly true in the inherently complex context of wicked problems. Thus, a common thread across many policy issues today is the growing conviction that public participation is a critical component in achieving any sort of progress with wicked policy problems (Selin, Schuett, and Carr 2000; Butler and Koontz 2005; Carroll et al. 2007).

Shannon (1992) asserts that the best way to manage an intractable problem is to provide an opportunity for community members to mutually exchange information and knowledge, and deliberate to arrive at the "solution" acceptable to the majority. Fischer (1993, 182) contends that "contrary to technocratic expectations," the right approach to dealing with wicked problems is "more—rather than less—citizen participation." In fact, he argues, "problem-solving in the case of wicked problems may literally depend on such collaborative methodological innovation." Dryzek (2000, 173) also supports this argument, asserting that "discursive democracy may be the most effective political means currently available to solve complex social problems, because it provides a means for coherent integration of the variety of different perspectives that are the hallmark of complexity." In the context of environmental management, Fiorino (2000, 540) argues that, "indeed, the themes of cooperation, participation, integration, ecosystems, and regional/local problem solving define a vision for a new era of environmental management, a vision that is emerging almost case by case across the country."

Analysts and decision makers are becoming increasingly aware of the disparity in how people perceive risk, which is contributing to the shift from traditional analysis to participatory decision making. As noted in chapter 2, it is especially important to acknowledge different risk perceptions since they are often at the root of controversy (Gray 1989; Slovic 2000). For example, government agencies frequently adopt a technical view of a given issue, which often leads to a different understanding of possible risks than the public's social or economic sensitivities (Gray 1989). A review of three case studies of fuels management and public participation found that experts and the public tend to emphasize uncontrollable factors when considering the causes of wildfire; emotional responses play a large part in judgments about wildfire risks; and "preferences for risk management options are remarkably malleable in response to even slight shifts in framing" (Arvai et al. 2006, 173). This different understanding frequently generates controversy over the accuracy of scientific reports and expert opinions, especially when "dueling scientists" disagree about conclusions drawn from the same information—and even about how much and what type of information is necessary to reach a sound decision. After twenty-five years of researching risk perceptions, Slovic (2000) concludes that a low level of technical understanding coupled with a high level of uncertainty and risk precipitates an increased public perception that there is potential for mistakes with devastating consequences. Given these findings, it may not be surprising that the public is significantly more likely to trust a university scientist's assertion than a government official's (Johnson and Scicchitano 2000).

Gray (1989, 253) contends that "in order to reach agreement on the acceptability of technical input, the parties will need to agree on the underlying value premise." This agreement can be accomplished only through a discursive process of mutual learning that allows all stakeholders to gain a better understanding of both their own position and others'. Findings such as these led the Committee on Risk Characterization of the US National Research Council to strongly advocate a decision process for situations involving risk that integrates both technical analysis and broadly based public participation (National Research Council 1996).

There are still practitioners who question the value added by participatory efforts in what many see as highly technical and scientific issues (Petts and Brooks 2006; Chilvers 2008). Nonetheless, many researchers, analysts, and policy makers seem to agree that participation in some form is now accepted practice in environmental management, and substantial literature has emerged on the design and evaluation of participatory efforts. Most of this literature focuses on participation per se and does not explore how such participation can or should work in the context of wicked problems or in processes that attempt to link ongoing scientific analysis and participation in what the National Research Council has termed an "analytic-deliberative" process (National Research Council 1996). For example, Chilvers (2008) notes that the public's role in scientifically framed issues remains unclear and that there are a number of important questions about this role, including the following:

- When, and to what extent, should the public be engaged in science and technology appraisals?
- How is contested knowledge best represented, communicated, and translated for participants in deliberative fora?
- Is mutual understanding attainable given large epistemic inequalities between the public and specialists?
- How should analytic-deliberative processes be structured to make space for cultural forms of rationality, instead of simply reifying instrumental rationality?
- To what extent should appraisal processes "open up" or "close down" wider policy discourses and uncertainties/indeterminacies? (Chilvers 2008, 157)

To explore these issues, we will first recapitulate the consensus view on how high-quality participatory processes should be designed. We will then consider this design template as it might apply to wicked problems, with the intent of developing criteria for the design of participatory processes for wicked problems.

The Design of Participatory Processes

As noted, there is now a substantial consensus among analysts and policy makers that addressing wicked problems requires a greater level of public participation than has been the case in traditional analytic approaches (Rittel and Webber 1973; Fischer 1993; Haight and Ginger 2000; Steelman 2001). Increasing calls for expanded public participation have been attributed to many factors, including a growing distrust of public institutions and officials (Fischer 1993; Thomas 1995); increased legislative requirements for public participation (Thomas 1995; Haight and Ginger 2000); the complexity and uncertainty of contemporary problems (Fischer 1993); different perceptions of risk (Krimsky and Golding 1992; National Research Council 1996; Slovic 2000); and a common recognition that decisions are never purely scientific—that politics and social values are inherent in all administrative decisions (Durning 1993; Fischer 1993; DeLeon 1995; Wondolleck 1988).

Despite widespread acceptance of the need for participatory processes, public participation has received mixed reviews from practitioners and evaluators. On the positive side, it has been credited with the ability to facilitate

- trust, legitimacy, and mutual learning by incorporating public values into decisions (Pateman 1970; Fischer 1993; Beierle and Konisky 1999; Wondolleck and Yaffee 2000; Daniels and Walker 2001; Beierle and Cayford 2002; Mascarenhas and Scarce 2004);
- leveraging public resources (Selin et al. 2007);
- conflict resolution among divergent interests (Gericke, Sullivan, and Wellman 1992; Beierle and Cayford 2002);
- better substantive decision quality based on a broader range of interests, debate, and innovation (Gray 1989; Sample 1993; DeLeon 1995);
- increased decision commitment and acceptance (Pateman 1970; Gray 1989; Sample 1993);
- positive cumulative partnerships to enhance future policy endeavors and create goodwill (Gray 1989; Gericke, Sullivan, and Wellman 1992; Beierle 2002); and
- stronger democracy built on a genuine attempt at inclusiveness (DeLeon 1995).

Conversely, it has also been argued that lay populations lack the technical knowledge and skills to seriously engage scientific issues (Petts and Brooks 2006). Broad participation has been described as "tyrannical" (Cooke and Kothari 2001) and is often associated with

- intensive resource commitments (e.g., money, time, and human capital);
- prolonged decision making (Sample 1993);
- decreased stakeholder trust;
- a perceived loss of agency or manager control (Gray 1985; Selin, Schuett, and Carr 1997);
- reduced decision quality;
- increased controversy and conflict, including co-opting by powerful interest groups (Gray 1985; Selin, Schuett, and Carr 1997); and
- diminished likelihood of successful outcome(s) (Steelman 2001).

The Design of Successful Participation

If effective public participation is essential in addressing wicked problems such as those presented in managing complex ecosystems in a highly charged political environment, stakeholders must incorporate the design considerations important in successful participatory efforts. For example, trust is often cited as one essential element in collective decision making, as it is necessary to foster cooperation and reduce individual opportunism (DeLeon 1995; Mitchell-Banks 2006; Ostergren et al. 2006). Conversely, lack of trust can hinder a successful outcome (Fischer 1993; Shindler, Brunson, and Stankey 2002). Thus, the higher the level of trust between agency and participants, the greater the likelihood of success.

Not surprisingly there has emerged a large literature on how participatory processes should be designed and evaluated. Reed's review of recent literature develops a list of best practices in participatory processes (Reed 2008; see also Webler and Tuler 2006; Chilvers 2008):

- Stakeholder participation needs to be underpinned by a philosophy that emphasizes empowerment, equity, trust, and learning.
- Where relevant, stakeholder participation should be considered as early as possible and throughout the process.
- Relevant stakeholders need to be analyzed and represented systematically.
- Clear objectives for the participatory process need to be agreed among stakeholders at the outset.
- Methods should be selected and tailored to the decision-making context, considering the objectives, type of participants, and appropriate level of engagement.
- Highly skilled facilitation is essential.
- Local and scientific knowledge should be integrated.
- Participation needs to be institutionalized.

As Reed (2008) notes, a relatively small number of individuals have made any effort to investigate the validity of either the positive or negative claims for participation, but after a careful review of the available literature, he concludes there is some evidence that greater participation contributes to better decisions. However, he adds a strong caveat that "the quality of a decision is strongly dependent on the quality of the process that leads to it" (Reed 2008, 2421).

A Proposed Approach to Participation for Wicked Problems

In this section, we draw on the insights gleaned from the best available thinking about participatory processes and then sketch the requirements for such processes that involve wicked environmental problems. To understand what a *quality process* is in the context of wicked problems, designers of successful participatory efforts need to answer three key questions: (1) What are the agency's objectives in seeking stakeholder input? (2) How and how much will participants be involved? (3) Where will the process occur in the cycle of policy development?

One approach for describing the objectives of participatory processes is reflected in Webler's identification of the Habermasian objectives of fairness and competence (Webler 1995). In this sense, fairness means both that the full range of relevant stakeholder views are considered and that power relationships between participants are somehow equalized. Competence here incorporates the view that the process should arrive at an improved decision (Renn, Webler, and Wiedemann 1995; Webler 1995; Reed 2008).

We can also learn something about objectives in considering the nature of public engagement. Rowe and Frewer (2000) note three types of stakeholder engagement in their assessment of participatory methods. Both "communication" (disseminating information to participants) and "consultation" (gathering information from participants) represent a one-way communication flow. Rowe and Frewer term two-way communication as "participation," which is much closer to what Renn calls "deliberation." Renn (2006) argues that public deliberation is important on three grounds. First, deliberative processes are necessary to define the role and relative importance of scientifically derived ("systematic" is Renn's term) knowledge versus local experience and more idiosyncratic knowledge. Second, deliberation is needed to find the most appropriate way to deal with uncertainty and to set efficient and fair trade-offs between potential overprotection and underprotection in the face of uncertain outcomes. Finally, Renn argues that deliberation is needed to address the wider concerns of the affected groups and the public at large.

Stirling (2006, 96) cites a "useful—if imperfect—practical heuristic distinction" by Fiorino (1990) in making the argument that participatory processes are undertaken from three different perspectives. First, greater participation may be attempted from *normative* considerations of democratic principle. This is the Habermasian view that contemporary societies should as a matter of principle engage as large a constituency as possible in making environmental decisions. Alternatively, we might have *substan-*

tive reasons for combining analysis and participation. From this perspective, participation holds the promise of increasing the breadth and depth of the information available to decision makers, and can thereby improve the quality of decisions made. Finally, greater participation can be undertaken in an effort to enhance public trust and the credibility of decisions reached. Stirling (2008) argues this *instrumental* imperative is relatively neglected in the academic literature, but is nonetheless real and relevant. We note that these motivations for seeking stakeholder input are not mutually exclusive. An organization may have multiple reasons for initiating a public input process, though it is quite conceivable that one of the motivations may dominate in a given case.

Understanding the nature and extent of participant involvement is also well represented in the typologies of participation. One of the first such typologies was Arnstein's ladder of participation (Arstein 1969). Reed (2008) provides an excellent summary of more recent variations on the Arnstein ladder, which generally amount to different labels for the ladder rungs. For example, Lynam and colleagues (2007) describe a continuum of engagement for participation divided into three "classes":

- diagnostic and informing methods that extract knowledge, values, or preferences from a target group to understand local issues more effectively and include them in a decision-making process;
- colearning methods in which the perspectives of all groups change as a result of the process, but the information generated is then supplied to a decision-making process; and
- comanagement methods in which all the actors involved are learning and are included in the decision-making process.

One shortcoming of most of the "ladder" approaches is the implication that higher rungs on the ladder are preferred to lower rungs (Arstein 1969). For example, advocates generally see comanagement as superior to colearning, and colearning as superior to diagnostic/informative methods. But as Reed (2008) notes, different levels of engagement are likely to be appropriate in different contexts, depending in part on the objectives of the process (see Tippett, Handley, and Ravetz 2007 for one example).

Walters, Aydelotte, and Miller (2000) argue that analysts and decision makers need to understand the stages of policy development in order to set appropriate expectations for public participation. Understood from the traditional policy analytic framework, it makes little sense to design a participation process intended to measure public opinion on specific alternatives if the problem is yet to be defined. At the same time, a process intended to educate the public and legitimize the selection of a final alternative is likely to create great frustration for all involved if the public expects to share their perspectives regarding how the problem should be defined or to offer new proposals for problem resolution. The challenge of understanding the policy development cycle is particularly important for wicked problems.

Increased public participation is not the answer to every policy decision. In some cases, the public expects administrative agencies to make decisions without any input, accepting their legal authority or expertise (Sample 1993). However, for wicked problems there is a substantial consensus across a number of disciplines that broadly based participation from all stakeholders in the decision-making process is essential (Fischer 1993).

As we noted in chapter 2, wicked problems have no definitive problem formulation, no stopping rule that indicates when the problem is "solved," no test for a solution, and no well-defined set of alternatives. If these conditions hold, as they often do in environmental decision settings, the stages of policy development identified by Walters, Aydelotte, and Miller (2000) must be modified. Their description includes the standard, policy analytic steps of defining the problem, identifying decision criteria, generating alternatives, evaluating the alternatives using the criteria, and recommending a course of action. But if it is not reasonable to expect a single definition of the problem, it makes little sense to engage in a process intended to generate a consensus on the problem definition. Further, since there is no reason to believe that wicked problems can be *solved*, processes that create expectations that a solution will be identified are also counterproductive. Because there is often not even agreement on a shared language or a conceptual framework, comparisons when no common basis for comparison exists become impossible. Faced with the kind of radical uncertainty inherent in wicked environmental problems, the impossibility of reaching widespread agreement on even approximate likely outcomes precludes arguments based on "the interest of everyone affected." In fact, faced with a wicked problem, even determining a stakeholder's interest is not straightforward (Pellizzoni 2003).

But the problem's nature is not the only factor that may result in a call for expanded participation. Table 6.1 summarizes lists provided by Gray (1989) and Selin and Chavez (1995). In the table, factors are grouped into four broader classifications relating to problem characteristics, stakeholder characteristics, the history of the decision, and the particular decision context. The more characteristics from table 6.1 are present in a given decision situation, the greater the apparent need for carefully designed and rich public participation.

TABLE 6.1. Factors calling for greater public participation (Gray 1989; Selin and Chavez 1995)

Problem/issue characteristics	Both the problem's definition and solution are illusory (i.e., wicked problems).
	The problem/issue involves uncertainty and risk.
Stakeholder characteristics	Various stakeholders have an interest in a problem and cannot achieve a unilateral solution.
	Stakeholders are not easily identifiable in advance or well organized.
	There is possible disparity in stakeholder power and available resources.
	Stakeholders have various levels of expertise and knowledge.
	Stakeholders have different values and interests that have led to adversarial positions on other occasions.
	A common goal exists between different stakeholders that requires a collaborative effort to fulfill.
	A network exists, such as a chamber of commerce or community organization, where stakeholders have already forged alliances.
History of the decision	Incremental or unilateral problem resolutions have not proven successful.
	Existing methodologies to address the issue have not been successful, and may even have compounded the problem.
Decision context	A crisis exists where policy action has been paralyzed by adversarial positioning, by legal delay tactics, or where there is immediate danger, such as the need to protect endangered species or ecosystems.
	A third party is intervening who convenes the stakeholders as a neutral mediator.
	There is a legal mandate handed down by the legislature or the courts, such as with the National Forest Management Act.
	A charismatic leader's enthusiasm and vision persuades others to join in the participatory effort.
	Incentives for participation exist, such as matching funds for stakeholders to participate.

But to say that there are multiple definitions of a problematic situation is not to say that all characterizations are equally viable. Likewise, to say there are conflicting values in play does not legitimize all preferences and goals. Perhaps most important, conflicting views on the nature of the problem and what the policy goals should be does not preclude broad agreement on what action should be taken next. The issue then is how to answer the three key questions about participation listed earlier most appropriately in the context of a wicked environmental problem.

According to participatory theory, resolving complex problems in practice requires approaches that are (1) acceptable to all stakeholders, (2) practical to implement, (3) technically feasible, (4) economically sustainable, and (5) politically achievable.

Participatory processes often lead stakeholders to *temporarily* develop ad hoc associations that come together to address specific issues. Attaining the necessary levels of trust and understanding to confront complex policy dilemmas requires *sustained* attention and involvement from the earliest stages. To establish this essential participatory infrastructure, the National Research Council (1996) proposes relying on developing "learning networks" (Senge 1990; Stubbs and Lemon 2001) of stakeholders to create a cooperative decision-making environment in which trust, understanding, and mutual reliance develop over time.

In the learning network process, analysis and deliberation build on each other in iterative cycles. This analytic-deliberative approach requires engagement by both scientists and public stakeholders. Analysis allows participants to develop and draw from a common, scientific knowledge base that informs deliberations. Deliberation in turn allows a policy consensus to emerge that is both socially and scientifically acceptable.

Unfortunately this process is not a panacea. Since participatory policy analysis democratically integrates various types of information—quantitative and qualitative, analytical and perceptual, and objective and subjective—it can be highly demanding of social resources, including time, money, and stakeholder commitment. As our review suggests, it is quite easy for such processes to get derailed, and there are many more instances of failed processes than of successful ones.

Several aspects of large-scale wicked problems make successful participatory processes problematic. First, both the public and agency decision makers tend to seek optimal scientific, if not political, solutions. But when confronted with a wicked problem, such a search is futile. In the context of wicked problems, therefore, searching for an optimal solution should give way to efforts that strive more realistically for mutual "satisficing" (Simon 1997, 295). To satisfice is to search for broadly acceptable and implementable solutions, rather than for optimal solutions that may be difficult to implement (Simon 1957). Again, optimal solutions generally either do not exist or cannot be identified in advance for wicked problems.

Second, as noted in chapter 2, there is the issue of scale itself. Successful participation efforts have almost always been local in nature (Shindler and Cheek 1999). Indeed, Fiorino (2000) sees this as a defining characteristic of the new era of environmental management. Yet some environmental problems require broader perspectives and actions at a larger scale. And all of a wicked problem's most difficult aspects are accentuated when the scale is increased.

Consider for example the national forests in California's Sierra Nevada, which comprise more than eleven million acres. The Sierra Nevada Forest Plan Amendment process has incorporated numerous opportunities for public engagement, and relevant documents repeatedly refer to the importance of *transparency* and *public involvement*. While retaining ultimate decision-making responsibility as required by law, over the past twenty years Forest Service administrators have strengthened opportunities for concerned citizens and stakeholder groups to participate in developing alternative forest management plans and to comment on alternatives under consideration. And yet controversy persists as evidenced by continuing appeals and litigations of the original decision and projects that arise from that decision. What more can the agency do?

To begin our answer, it is helpful to first identify who the participants should be. Pellizzoni (2003) discusses the distinction made by some writers between public participation and stakeholder participation. In this view, the public is defined as nonorganized lay citizens, while stakeholders include interested and affected individuals and groups (Beierle and Cayford 2002). The utility of the distinction, Pellizzoni argues, lies in the fact that successful participatory processes will differ depending on whether one is engaging the public or stakeholders. The public at large has what Pellizzoni calls "normative competence" in that they hold relevant opinions, preferences, principles, and values, which the participatory process should attempt to elicit. But organizers should not expect participants to provide new knowledge on the issues.

Stakeholders on the other hand have normative competence in this sense, but they are expected to also have cognitive competence. Stakeholders are individuals, groups, and organizations—including public agencies and government representatives—who have an interest in an issue or problem (Gray 1989; Selin and Chavez 1995). While some agencies identify stakeholders as individuals and groups external to the agency, our usage includes both internal and external interested parties. Further, stakeholders can define the issue, or the issue can define the stakeholders (Stubbs and Lemon 2001). One of the major challenges of participatory efforts can be the need to identify groups of stakeholders and bring these "clusters of shared interests and concerns" (Stubbs and Lemon 2001, 323) together in a learning-conducive environment (Gray 1989; Beierle and Konisky 1999; Junker, Buchecker, and Muller-Boker 2007). Only then, it is argued, can stakeholders transcend their usual boundaries and belief systems to gain greater awareness of other stakeholders' perspectives and sensitivities, and to construct a mutually acceptable understanding of the wicked problem at hand (for example, see Beierle and Konisky 1999; Selin et al. 2007). Stakeholders should be engaged both because of their normative competence and because of their possible contribution to a better understanding of the problem and potential solutions. Thus, what is expected in stakeholder processes goes beyond simple representation. As Pellizzoni notes, democratic systems already offer many other opportunities for interest-based conflict resolution. What is sought in addition in a stakeholder process is a positional insight into a problem "derived from their looking at it from a specific professional or social viewpoint, and often for a long time" (Pellizzoni 2003, 200).

One of the features of an analytic-deliberative process, as advocated by the National Research Council (1996), is that experts and stakeholders interact regularly. The nature of this engagement can be grouped into four categories: discovery, deliberation, aggregation, and evaluation. The first two qualify as opinion processes in Pellizzoni's categorization, while the last two are position processes. Our four categories bear a strong familial resemblance to the five major processes of participation identified by Tippett, Handley, and Ravetz (2007) but are specifically tailored to the demands of participatory processes in the context of wicked problems.

Discovery processes are intended to elicit, either from the public or from stakeholders, values, concerns, aspirations, and insights into the environmental situation being considered. These processes should be both broadly based and inclusive, and intended to identify issues, potential criteria, possible alternative courses of action, and the value positions of both the public and relevant stakeholder groups. They occur either early on in the development of a course of action or in response to an adverse evaluation of past actions.

A deliberative process takes the learning from prior discovery processes and links it to the best available science. Using the concerns, aspirations, and values elicited during the discovery process to guide the scientific efforts, the science should explore the technical feasibility and potential implications of proposed alternatives, espoused ecological goals, and conflicting values. To be most productive, this should be an iterative process in which science and stakeholders engage in mutual learning and deliberation. As stakeholders more fully understand the potential scientific implications of particular actions, their value positions and goals will undoubtedly change, or new and unanticipated concerns may emerge. The ultimate objective of this deliberative process is to identify a small set of specific actions or next steps.

The criteria for a course of action to be included in this small set of options should be that it is broadly supported among stakeholders. It is expected that there will be serious concerns from some groups about certain options. But each included option should receive substantial support from at least a large minority of stakeholders, and should respond to their concerns and priorities. The objective of deliberative efforts should be to capture the essential decision trade-offs that must be made by identifying the actions deemed scientifically defensible and most acceptable to various stakeholders.

Once process managers have identified scientifically sound and likely acceptable options, the nature of public engagement changes. The process moves from collecting opinions to aggregating preferences. Aggregation processes are about negotiation and trade-offs, and the objective is at least a majority willingness to accept a given course of action, selected from the set of options identified in the deliberation process. If no such majority is obtained, the results along with the concerns and scientific information available are forwarded to the decision maker(s), for something must be decided and some action will be taken even if it is to continue the status quo.

But even after the decision is taken and implemented, there is an ongoing role for stakeholder participation in evaluating the implementation and outcomes. Wicked problems are not solved, they are managed. And the best management of such problems requires that stakeholders be involved in an ongoing process that may grow out of adaptive management initiatives, which will certainly provide feedback and assessment of changing trends and patterns. The issue to be confronted in evaluative processes is whether the combination of management actions, current field conditions and the best available scientific understanding of trends combine to yield a broadly accepted result. If so, current practices will continue. If not, a new cycle of at least deliberation and perhaps discovery will be initiated.

Throughout this cycle of discovery, deliberation, aggregation, and

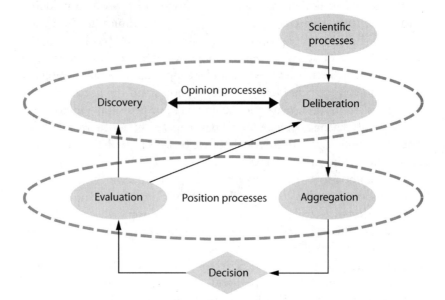

FIGURE 6.1. A conceptual model of discovery, deliberation, aggregation, and evaluation.

evaluation, the objective is not agreement on values, problem definitions, or even facts. The objective is to agree on what should be done next. As Pellizzoni (2003, 210) puts it, "The matter becomes no longer to find a common reason, but to reach agreement on a practice; no longer to go upwards, to abstractions, but downwards, to concrete solutions for concrete and circumscribed situations." Figure 6.1 summarizes the conceptual model just described.

Design Requirements

In the context of wicked problems, the National Research Council (1996) has called for analytic-deliberative processes. But past efforts leave it less than clear how participation fits into such processes. Chilvers (2008) argues that stakeholders generally agree on the effectiveness criteria for participatory processes, but they understand less how participation should work in an analytic-deliberative process. Chilvers identifies fourteen principles that should guide participation linked to science in an ongoing deliberative process. Chilvers' first two principles are particularly relevant for discovery processes. In an analytic-deliberative process focused on discovery, the public and stakeholders should be actively engaged in the earliest framing stage to (1) identify alternative formulations of the problem, alternative courses of action, and acceptability criteria; and (2) shape and guide scientific analysis conducted throughout the process (Chilvers 2008).

The remaining principles Chilvers articulates relate to deliberative processes. These include principles relating to public engagement, the role of scientific analysis, access to information and expertise, and the quality of deliberative efforts. Chilvers argues that in an analytic-deliberative process focused on deliberation, the public and stakeholders should be actively engaged in scientific assessment and evaluation where they demand to do so or where science supporting the decision process is particularly contentious or uncertain.

Scientific analysis in the analytic-deliberative process should support deliberation and be accessible, relevant, and usable to participants within the process; be responsive to the needs, issues, and concerns expressed by participants in an iterative way; and be transparent to participants within the process and make underlying uncertainties and assumptions explicit.

In relation to *access to information* and specialist expertise, the following is advised:

- Information provided should be appropriate, meaningful, and understandable from the perspective of those participating.
- Information provided within the process should faithfully represent the range and diversity of views that exist on the issue being considered.
- Information provided within the process should be responsive to the needs of participants.
- Participants within the process should have access to specialist expertise and control over who provides this assistance.

Deliberation conducted within the process should do the following:

- Ensure a highly interactive, symmetrical, and critical relationship between participants and specialists.
- Emphasize diversity and difference through representing alternative viewpoints, exploring uncertainties, and exposing underlying assumptions.
- Allow enough time for participants to become informed and develop competent understandings.

- Ensure that facilitators have adequate substantive understanding of the issues while remaining independent and impartial as to the outcomes of the process (Chilvers 2008).

The primary objective of deliberation in the framework presented here is not to arrive at a decision or a recommended course of action. Rather, the objective is to explore as richly as possible the implications of alternative courses of action. Such an exploration should encompass an understanding of stakeholder attitudes and reasons for those attitudes; the decision elements that are of greatest concern to stakeholders; the widest possible range of alternative courses of action; and the environmental, social, and economic implications of decision alternatives, with particular focus on those decision elements of greatest concern to stakeholders.

There is no expectation that unanimity will emerge from the deliberative process. Indeed, if it does, there was no wicked problem to begin with. Expect that during the course of deliberation, people's value positions will change, everyone's scientific understanding will improve, and greater clarity on exactly where and why people disagree will emerge. And the end product of the deliberative process should be a (relatively) small set of decision alternatives, specific actions in specific locations over specific time frames that capture the range of concerns and desires of participants. Each alternative will also involve a current estimate of likely outcomes, but given the uncertainties of the underlying processes, what matters most are the actions, locations, and time frames.

Once this working set of alternatives is generated, the nature of the task shifts to an aggregation of public and stakeholder preferences. The difficulty of this shift should not be underestimated. As figure 6.1 suggests, there is a fundamental shift at this stage from an opinion process to a position process. Discovery and deliberation, even evaluation, are about "opening up" the analytic-deliberative process, to use Stirling's term (Stirling 2008). Aggregation is about "closing down" the process and narrowing the range of alternatives. Given the nature of wicked problems, this inevitably is a contentious process. It is at this stage that deliberative democracy confronts social choice theory.

The essential optimism of advocates of deliberative processes is that through open and active deliberation, conflicts will be mediated or transformed, common ground will be found, and a consensus on the best course of action will emerge. But by the very nature of wicked problems, such an outcome is highly unlikely. Social choice theorists on the other hand argue that any aggregation of preferences short of unanimity is essentially arbitrary and is as much a function of how the aggregation is done as it is of the underlying preferences (Riker 1982). Such a position seriously undermines the whole notion of the legitimacy of deliberative efforts (Knight and Johnson 1994). While there have been recent attempts to reconcile deliberative democratic theory and social choice theory (e.g., Dryzek and List 2003), for most writers there remains a substantial chasm between the two views.

And this is more than a mere academic squabble. If participatory efforts are truly essential to the management of wicked environmental problems, there must be a sound and defensible method for summarizing, or aggregating, the views of participants. But if the aggregated preferences are a product of the aggregation method as much as the participant's preferences, how should any resulting ranking of decision alternatives be understood? There is no simple way out of this dilemma.

A Possible Option

Our proposal is that stakeholders and public representatives first clearly separate the deliberative process from the aggregation process when discussing wicked problems. Again, we would argue that the appropriate result of the deliberative process should not be a recommendation or decision, but it should be a set of detailed alternatives that captures the range of stakeholder preferences for actions across the technically feasible and environmentally sustainable range of likely outcomes. The objective of the aggregation process is then to characterize the level of support that each alternative is likely to have among the public and stakeholders. The information gathered from the deliberative process can be used to develop support profiles for the alternatives under consideration as we demonstrate in the next several chapters. To be sure, there is an element of arbitrariness in any such characterization in that the resulting profiles will in part be influenced by the methods used to create it. We note here simply that in the course of the discovery and deliberation processes, rich and sophisticated data on stakeholder preferences will be gathered and refined. This information can be used in the aggregation process to anticipate the level of support for the alternatives under consideration.

At this writing it appears that the best method to use for converting preference information into support profiles is a variation of what Brams and Sanver (2009) call "preference approval voting." To implement this approach, two types of information must be collected as part of the deliberative process:

- an indication of whether or not a given decision scenario (i.e., a combination of actions, locations, and timing linked to potential short- and long-term outcomes) is acceptable, and
- a preference ordering of the decision scenarios.

Note that these two types of information are quite different. A given individual may prefer scenario A to scenario B, but may find one, both, or neither acceptable. If an individual finds a particular scenario acceptable, then we term that judgment as an *approval vote* for that option. Thus, if there are ten options under consideration, and a given participant finds that four would represent acceptable combinations of actions, locations, timing, and outcomes, while the other six are unacceptable for whatever reason, the four acceptable scenarios would receive one approval vote each. The six unacceptable scenarios would receive no approval votes from that participant. The preference ordering provided by the participant would reveal which of the four acceptable scenarios is the participant's most preferred combination, second most preferred, and so on.

With this information from participants, decision options can be ordered based on the following two rules, adapted from Brams and Sanver (2009):

- If no alternative or exactly one alternative receives a majority of approval votes, the alternatives are ranked based on the number of approval votes received.
- If two or more alternatives receive a majority of approval votes, then use the preference ordering information to rank the alternatives that received at least a majority of approval votes.

To be clear on how these two rules interact, consider the following simple example. Suppose there are three decision alternatives under consideration labeled A, B, and C. Now suppose there are four stakeholder groups participating (labeled J, K, L, and M), and for the sake of simplicity, assume they each consist of twenty-five participants. Assume further that each group has the preferences shown in table 6.2, where "A > B" is read "A is preferred to B."

Under the aggregation rules, we first count the number of approval votes each alternative receives. Again, an approval vote is an indication that an alternative is considered acceptable to that group. So for group L, option B is preferred to option A, and A is preferred to C. Using the information in the column "Acceptable options (1)," group L finds only option B accept-

TABLE 6.2. Preference approval voting example (Brams and Sanver 2009)

Stakeholder group	Preference ordering	Acceptable options (1)	Acceptable options (2)
J	A > B > C	A & B	A & B
K	A > B > C	A	A
L	B > A > C	B	A & B
M	C > B > A	B & C	B & C

able. From the table we can see that alternative A is considered acceptable to groups J and K, and therefore receives fifty approval votes. Alternative B is considered acceptable by groups J, L, and M, and so receives seventy-five approval votes, while alternative C is acceptable only to group M and thus receives twenty-five approval votes. Since only alternative B received a majority of the approval votes, the aggregate ranking of alternatives would be B > A > C.

Now suppose that the acceptability ratings are as shown in the column labeled "Acceptable options (2)." In this case, group L finds both B and A acceptable, even though B is still preferred to A. Counting the approval votes yields seventy-five votes for both A and B, and twenty-five votes for C. Since both A and B received a majority of approval votes, under the second rule we use the preference ordering to rank these alternatives. Since A is preferred to B by two groups (J and K), while B is preferred to A by only one group (L), the final ordering would be A > B > C. This example shows both how the aggregation rules are applied and how changes in approval ratings might change the final rankings. While this shift in the final ordering may seem counterintuitive since preference orders did not change, a careful review of the table shows that there are no surprises. In the first instance three (equally weighted) groups found option B acceptable. It was the first choice of only one of these groups, but it was acceptable to two others. In the second case, both options A and B were acceptable to a majority of participants, and a majority of that majority preferred option A to option B.

Thus the outcome of the aggregation process is a ranking of the alternatives based on the preferences and acceptability ratings given by the participants in the deliberative process. This information and supporting analysis is then forwarded to the decision maker(s). The decision selected may or may not be in accordance with the rankings from the stakeholder participants, but it will certainly be easier to defend a course of action that coincides with the output from the deliberative and aggregation processes.

Once the appropriate entity implements the decision, they start an evaluative and monitoring process to track the success of the selected strategy. We anticipate that this evaluation process will review field data trends and updated science models, and will recommend either continuation of the selected course, revisiting the deliberative process or if necessary reinitiating a discovery process.

Embedding Participation in Formal Processes

There is nothing in practice or literature to suggest that well-designed and well-executed participatory processes will necessarily be considered successful either in terms of enhanced public support for the ultimate decision or in some instances influence on the final decision (Webler 1995; Rowe et al. 2008). While there may be broad acceptance of public participation among policy elites and academics, there is less acceptance among substantive experts and practitioners (Petts and Brooks 2006). There are a few examples of positive results from participatory processes (see the examples listed in Dietz, Ostrom, and Stern 2003), but many more efforts fail to meet expectations.

The description of the efforts by the Forest Service to engage the public exemplifies the fact that throughout the development of the environmental impact statement (USDA Forest Service 2001a) and the review effort (USDA Forest Service 2003), the Forest Service made a concerted effort to solicit public participation and incorporate public priorities in their decision process. While the quality of facilitation across the many meetings and forums inevitably varied, we cannot assess whether the Forest Service's participation efforts incorporated all the desirable public participation attributes described here. Our research does suggest that the Forest Service employed the best available methods to enable and evaluate public participation. And yet the process resulted in the highest number of appeals on record at that time, and there is no indication that the degree of controversy declined as a result of the review process or the level of public participation.

It seems that although the best available public participation techniques and methods known at the time were used, efforts to both include and satisfy a substantial number of stakeholders fell short in the context of the Sierra Nevada Forest Plan Amendment process. We suggest that this is not the Forest Service's failure, but rather an indication that traditional approaches to public participation do not work well in large-scale wicked problem contexts. Historically, *public participation* has been interpreted to mean providing opportunities for the public to comment on proposed federal actions that may impact the environment.

The National Environmental Policy Act and the National Forest Management Act make it clear that a federal official is responsible for deciding on a course of action. The decision maker cannot delegate this authority to any outside individual or group. Further, the decision maker must balance legal mandates and regulations, budget realities, public desires and expectations, and biological limitations in reaching a decision, guided by the values and risks at stake. This is especially difficult in the case of wicked problems where the science is uncertain and there is no consensus on public desires and values. It is also clear that the decision maker's personal values play a role in the final decision. In fact, according to the Forest Service, defining acceptable risks is part of the manager's job (USDA Forest Service 1997). Since there is no optimal solution to the Sierra Nevada management problem or other wicked problems, the greatest short-term management risks are related to decision processes rather than ecological outcomes. The best approach for dealing with such decisions is to develop a decision process that tightly integrates the best scientific analysis and the fullest possible public participation and deliberation.

Conclusions

The participatory approach we have outlined incorporates the key aspects of the Forest Service's approach under the requirements of the National Environmental Policy Act (the NEPA process) and the agency's own forest-planning processes. The identified issues determine the scale and nature of the analysis and decision, which drives the planning process. The suggested learning network process begins with identifying issues (see fig. 6.2), but also—at the outset—formally seeks to identify stakeholder values and preferences about these issues through a discovery process. This information could then be used as important input in developing a modeled set of alternatives that are environmentally, economically, and technically feasible and also reflect the range of stakeholder preferences and values. This step is motivated by the observation that there are generally a very large number of technically feasible potential solutions, not just the relatively small number usually displayed in a formal environmental impact statement. Further, since there are no perfect solutions to wicked problems, only those that are

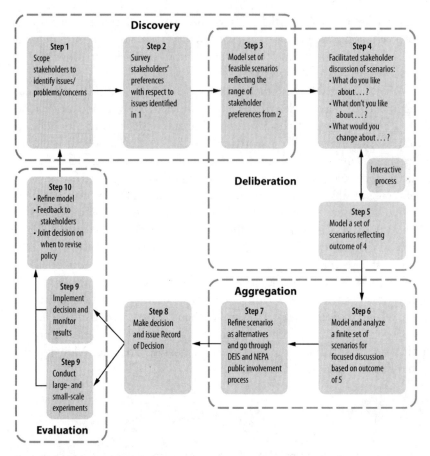

FIGURE 6.2. An adaptive, iterative, deliberative, analytical, participatory process for natural resource decision making in a wicked problem environment.

more satisfying or more useful than others, it will be important to find solutions that allow the agency to move forward in an adaptive management mode with a broad base of public support.

The output of this iterative, analytical process should be a well-defined, small set of feasible alternatives, perhaps three to eight, that best satisfy public preferences and values (Miller 1956). This set of alternatives could then be fully analyzed, subjected to the NEPA process and an aggregation process, and then decided upon. The decision would then be implemented, and a combination of monitoring results from implementation and small-scale experiments, jointly identified by stakeholders (managers, public interest groups, and scientists) to answer specific questions, could be used to provide feedback and evaluation to identify new issues and opportunities for further progress. One consideration that could foster understanding and the building of trust would be to use third-party or stakeholder monitoring.

This learning network process more explicitly models stakeholder preferences and uses these preferences to develop and assess initial alternatives. In the context of wicked problems, it is not reasonable to ask the public what should be done. By the nature of the problem neither the specialists nor the public would know the answer to that question, and neither would be able to recognize the best solution even if it were presented. There are appropriate questions on which to seek public input, but they depend on the stage of development of the analytical efforts—ask the public what they value and what their aspirations are for a given decision context. Science can then be used to explore the human, economic, and environmental implications of those values and aspirations. Public values and aspirations can be linked to the best available social and environmental science to report in a deliberative process on the implications of various value positions, the key decision trade-offs that must be considered, the areas of consensus and conflict among the various publics and stakeholders, and those decision alternatives that merit further development and debate.

Modeling preferences allows environmental managers to assess the values and preferences of a broad cross section of stakeholders. Explicitly linking those preferences to the science models used to generate alternatives samples a much broader range of alternatives and increases the likelihood of identifying alternatives that offer the greatest promise of being broadly acceptable. Decision makers are thus informed not only of the best available science but also of the multiple value positions held by the public and how the public is likely to respond to various decision alternatives both in the short term and the long term.

Chapter 7

A Proposed Adaptive, Deliberative Decision Process

At the end of chapter 6, we introduced a proposal for an approach to decision making in the wicked problem context that incorporates key elements of procedures required under the National Environmental Policy Act, known as the NEPA process (Council on Environmental Quality 1987, 2007), and the concept of *learning networks* as described by the National Research Council (1996). Figure 6.2, presented on page 126 of the previous chapter, depicts this approach.

In this chapter, we discuss this decision approach in greater detail. In doing so, we explain how it satisfies the requirements of the NEPA process and adopts the learning network concept. We also review how our novel techniques for formal elicitation and analysis of stakeholder preferences fit within the proposed approach. We close the chapter with a consideration of how the approach may be useful in the context of the wicked problem case studies considered throughout the book. Next, in chapters 8 and 9, we describe our preference elicitation and analysis techniques more comprehensively and report the results of a pilot test of the methods that we conducted during our research in the Sierra Nevada case.

The Decision Approach

The first three steps of the decision process illustrated in figure 6.2 constitute the discovery phase. In these steps the environmental management agency with responsibility for the decision at hand works to identify the stakeholders and draw out their concerns and preferences. The term *stakeholders* should be defined broadly. Stakeholders may include both individuals and groups representing larger interests, such as those of conservationists, recreationists, private sector firms, and so forth. These stakeholders may have limited means or may be sophisticated and well financed. They may have a local, national, or even international perspective. We suggest further that personnel within the management agency, including administrators and scientists, along with representatives of other relevant local, state, and federal agencies, should also be considered stakeholders, as should public officials representing their constituencies.

In our proposal, the agency during the discovery process identifies the stakeholders, identifies their key issues and concerns, systematically surveys their preferences regarding these issues and concerns, and then begins to develop the large set of feasible policy options that may reflect the wide range of preferences. Constraints to feasibility that may cause some options to be excluded at this stage could be scientific, technical, administrative, or budgetary. The process of identifying the constraints, and determining whether options should be rejected given these constraints, should also incorporate stakeholder preferences elicited during the discovery phase.

The Forest Service in the Sierra Nevada case, following commonly used agency procedures to satisfy NEPA requirements, implemented efforts to *scope* public concerns and seek public input. In general practice, however, agencies do not systematically survey stakeholder preferences as we recommend. Consequently conventional scoping processes typically result in a relatively small set of alternatives to undergo further evaluation. In the Sierra Nevada Forest Plan Amendment case, for example, eight management options were identified out of potentially thousands of possible scenarios. The discovery steps illustrated in figure 6.2 aim to provide a richer understanding of both stakeholder *perceptions* of the problem dimensions of greatest concern and stakeholder *preferences* for management strategies, including actions, timing, locations, and outcomes.

The elicitation of perceptions and preferences can be accomplished through various forms of outreach. These may include opportunities for public comment through mail, call-in numbers, or interactive websites; informational workshops and public service announcements prepared and delivered by agency personnel; and regular town hall meetings encouraging the participation of interested and affected parties. As described in chapter 6, however, many of these steps as currently implemented generally fall into the category of one-way communication—that is, they do not achieve the standards of full participation or deliberation delineated by scholars (Rowe and Frewer 2000; Renn 2006).

In combination with step 3 in figure 6.2, steps 4 and 5 move the process in the direction of genuine two-way communication. As noted in the figure, we categorize these three steps as the deliberation phase. While in the discovery phase the agency should work to gather ideas from as wide a range of stakeholders as possible, in practice a smaller number of individuals and representatives of groups will become more directly involved in the deliberation phase. For a productive learning network to emerge, these more engaged stakeholders would ideally reflect and represent a broad range of perspectives, have a long-term commitment to the effort, and have credibility with other individuals and groups who may be interested in, but not directly involved with, the agency's decision.

Step 4 of the deliberation phase centers on drawing out participants' responses to the set of options identified during the earlier systematic preference elicitation efforts as both technically feasible and potentially acceptable to a range of stakeholders. These deliberations are necessary for two reasons. First no formal preference elicitation and modeling effort can produce a comprehensive understanding of stakeholder values. Second, stakeholder value positions on various aspects of the problem under consideration are likely to change as participants explore the implications of those positions, taking into account the best available science and the best available models projecting outcomes from alternative management interventions. The deliberation phase enables precisely this type of exploration and learning, both about the technical aspects of the problem and about guiding values. The process will also help identify gaps in the knowledge base relevant for better technical understanding and afford stakeholders the opportunity to more fully understand the trade-offs inherent in their values and preferences.

In this way, the deliberative phase allows participants to narrow the often very large set of feasible policy options identified in step 3 to a smaller set of potentially acceptable and implementable options. The range of alternatives developed in step 3 should be analyzed only to the extent that they can be useful to stakeholders in further narrowing their choices in steps 4 and 5. Options considered unacceptable by most or all of the participants, either in terms of implementation strategies, including actions, timing, and location, or projected outcomes can be eliminated from consideration.

As indicated in the figure, steps 4 and 5 involve an iterative, interactive process. In step 4, participants engage in deliberative discussions to select policy alternatives of interest. In step 5, participants analyze these alternatives in concrete detail, assessing the costs, risks, and benefits of the associated implementation strategies and projected outcomes. The participation of scientists, modelers, and other technical experts from the agency in these steps is an important component of the approach. Through their engagement, all participants can be involved in establishing appropriate information standards and understanding the strengths and limitations of technical models.

Following this period of assessment, participants, through a process of mutual learning, reevaluate and modify the options under consideration. The iterative, interactive process followed in steps 4 and 5 allows participants ultimately to define a small set of alternatives, perhaps half a dozen, with the greatest potential for being feasible and implementable and for satisfying (or *satisficing*) public preferences and values.

In the aggregation phase, the agency again reaches out to the broader public both to report the results of the deliberation phase involving the engaged stakeholders and to obtain information on the preferred rankings of the identified alternatives. In steps 6 and 7, the potentially feasible, implementable, and satisficing options are developed more fully into scenarios. These scenarios in turn underlie the development of management alternatives that are formally presented and appraised in an environmental impact statement, as illustrated in step 7. The process of preparing environmental impact statements, a key requirement of the National Environmental Policy Act, includes further mandated steps for broad public comment.

The environmental impact statement process also leads to the development of sophisticated ecological models projecting the results of the management interventions under consideration. As discussed later in this chapter and in chapter 9, formal models of stakeholder preferences can be integrated with these ecological models in useful ways. With statistical techniques, it is possible to simulate stakeholder responses to various environmental outcomes predicted by the ecological models. In an ongoing learning network, new information derived from such simulations strengthens the deliberative phase, illustrated in steps 4 and 5 of figure 6.2, and may also prove useful in the aggregation phase, illustrated in steps 6 and 7. In the aggregation phase participants assess both the projected ecological impacts of each alternative under consideration, taking into account the uncertainties surrounding these projections, and the public preferences for the combinations of action, location, timing, and likely outcomes represented by the alternatives as expressed in the preference elicitation exercises.

Next, in step 8, the responsible agency administrator selects a policy alternative from among those assessed in the environmental impact statement, and this choice is formalized in a record of decision. In our approach, this step is followed by the evaluation phase, steps 9 and 10, centering on adaptive management. As discussed in chapter 5, we argue that adaptive management is an essential component of any response to the uncertainties and risks inherent in wicked problems.

In our decision approach, steps 9 and 10 incorporate adaptive management in ways that extend the adaptive component of the management process beyond ecological monitoring or scientific experimentation. The stakeholders who are engaged in the deliberative phase are also involved in the adaptive management implementation, thereby extending the adaptive approach to incorporate social and political aspects of the decision dilemma. Step 9 is divided into two options to accommodate the two types of adaptive management, passive and active, introduced in chapter 5 (Wilhere 2002). Passive adaptive management follows a process of implementation and monitoring, while active adaptive management establishes formal, controlled policy experimentation. The choice between the two types of adaptive management in any given case may depend on ecological, technical, or budgetary conditions. In any event, step 10 follows, in which the results of the monitoring or experimentation become input into subsequent rounds of the decision process. As discussed in chapter 2, a wicked problem by definition has no identifiable stopping rule. As some aspects of the problem become less salient, others will become more problematic and divisive. In this context, participants, including agency staff and public stakeholders, need to engage in a process of continuous mutual learning informed by outcomes of prior management actions, emerging scientific knowledge, and evolving public priorities. The process illustrated in figure 6.2 continues in repeated iterations as long as the uncertainty and conflict remain.

The NEPA Process

For our proposed decision approach to be practical, public managers must be able to apply it within common, real-world statutory and regulatory frameworks. The NEPA process is a widely used standard. The National Environmental Policy Act requires application of the NEPA process at the federal level in the United States, and many US states and more than eighty-five nations worldwide have adopted similar procedures (Caldwell 1998). Figure 7.1 presents a graphical depiction of the NEPA process (Council of Envi-

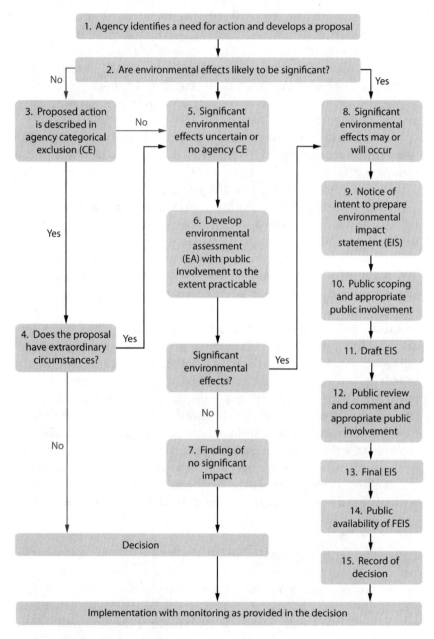

FIGURE 7.1. The NEPA process
Note: Adapted from the Council of Environmental Quality (2007)

ronmental Quality 2007, 8). In this section we discuss how our proposal incorporates the NEPA process, and thus can be applicable in many contexts.

In the United States, the NEPA process begins when a federal agency develops a proposal for action in response to an identified need, illustrated in step 1 of figure 7.1. In complicated or large-scale decisions, multiple agencies may be involved, including federal, state, and local entities. In such cases, the agency with primary responsibility becomes the lead agency, and the others are designated joint-lead agencies or cooperating agencies. In the next step, the agency assesses the likelihood of significant environmental effects from the proposed action. Three outcomes of this initial analysis are possible: (1) the proposed action is among a set of activities that have been predetermined not to have significant environmental effects; (2) the proposed action may lead to some environmental effects, but these impacts are unlikely to be significant, or there is uncertainty about whether they will be significant; or (3) the proposed action is likely or certain to lead to significant environmental effects. These three alternative initial outcomes are illustrated in figure 7.1 as steps 3, 5, and 8.

Step 3, the finding that an action is on a list qualifying it for exclusion, usually leads to a determination that further steps in the NEPA process do not apply. In these cases the agency can move toward making and implementing a decision without conducting environmental assessments or environmental impact statements. The "extraordinary circumstances" mentioned in step 4 refer to situations in which "endangered species, protected cultural sites, and wetlands" may be impacted (Council on Environmental Quality 2007, 11).

Step 5, the finding that there is uncertainty about environmental effects, leads to the requirement for an environmental assessment. An environmental assessment is conducted to determine whether the potential environmental effects of the proposed action may be significant—in which case a full environmental impact statement must be undertaken—or not significant—in which case the agency issues a formal finding to that effect, step 7. If the agency finds no significant impact, it may proceed to the steps of decision and implementation.

The process of conducting an environmental assessment includes some public involvement, but the participatory requirements are much more modest than those mandated when a full environmental impact statement is needed. Agencies also have considerable discretion as to the level of participation they pursue. For example, they may simply make the environmental assessment and the draft finding of no significant impact available to the public and interested parties before finalizing the process, or they may engage in scoping stakeholders and offering opportunities for public comment before proceeding.

Step 8, the finding of likely or certain environmental effects, leads to the requirement for a full environmental impact statement. This finding initiates a demanding and time-consuming set of procedures that includes comprehensive scientific study and extensive public engagement. An environmental impact statement may take several years to complete, and the final document may run to thousands of pages in length. As wicked environmental problems by definition entail at least the probability of significant environmental effects, the key steps in the NEPA process relevant for our discussion are illustrated in steps 9 through 15 in figure 7.1.

To begin the process, as indicated in step 9, the agency must formally publish a notice of intent to prepare an environmental impact statement. This notice describes the problem and alerts the public to the beginning of the scoping process. The scoping activities, step 10, are designed to clarify the roles and responsibilities of the participating agencies, identify stakeholders, establish lines of communications between agencies and stakeholders, determine issues of concern, enumerate gaps in the knowledge base, and set tentative timelines. Options for public communication and participation during the scoping process may include workshops, town hall meetings, video conferences, interactive websites, and so forth.

Next, in step 11, the agency works to prepare a draft environmental impact statement. This document draws on information gathered from the scoping process and from scientific studies conducted to fill gaps in the knowledge base. A central component of the draft statement is the presentation and appraisal of alternative policy options that could be selected to address the need initially identified by the agency. The purpose is to ensure the consideration of a range of feasible and reasonable alternatives, including the alternative of no action, so that potentially effective and efficient policy options are not overlooked. In the document, the full potential consequences of each alternative are presented, including "ecological, aesthetic, historical, cultural, economic, social, or health impacts, whether adverse or beneficial" (Council on Environmental Quality 2007, 17). In strengthening this attempt to clarify the costs, risks, and benefits of the alternatives in the present and over time, many experts are involved in collecting and analyzing data and developing models for forecasting ecological, social, and economic outcomes.

After the draft environmental impact statement is completed, a new round of public involvement begins, as indicated in step 12. During this period, which can continue for several months, stakeholders and interested parties, including those from the general public and from other government agencies, can comment on the technical background or on the proposed policy alternatives. Typically during the comment period the agency will conduct public hearings and other types of meetings and workshops to explain the document's content, answer questions, and invite responses.

Next, in steps 13 and 14, the agency prepares a final environmental impact statement. This revised document must include responses to comments received. This process of revision often requires substantial new effort, including further technical analyses, modifications of proposed alternatives, or even the consideration of new alternatives. After publication of the final environmental impact statement, and a required waiting period of at least thirty days, the responsible agency administrator makes a decision by selecting a policy alternative from those considered in the environmental impact statement. The formal record of decision accompanying this selection, step 15, explains and justifies the decision.

Our model, illustrated in figure 6.2, incorporates the NEPA process. Steps 1 and 2 in figure 6.2 provide opportunities for the scoping required in steps 9 and 10 of figure 7.1. Steps 7 and 8 in figure 6.2 integrate the full process of completing draft and final environmental impact statements and promulgating the record of decision. The key component our proposal adds that is not included in the NEPA process is richer and deeper deliberative participation by key engaged stakeholders, illustrated in steps 3 through 6 and 10 in figure 6.2. This deliberative, adaptive approach is designed to explore the range of available alternatives more widely and systematically than typically occurs during the process of preparing environmental impact statements under the NEPA process. In these steps, the process depicted in figure 6.2 leads to the emergence of learning networks, discussed in the next section, enhanced through the formal preference elicitation and analysis described in detail in chapters 8 and 9.

As noted in chapter 6, we found in our research on the Sierra Nevada case study that the Forest Service had gone well beyond the basic requirements of the NEPA process in its public participation efforts. The agency held numerous meetings throughout the region over an extended period. It also fostered a long-running connection with the most engaged stakeholders. Moreover we found that several of the policy alternatives presented in the draft and final environmental impact statements reflected the preferred options of key stakeholders, including public interest groups and other government agencies. Nevertheless, the process failed to reduce the conflicts. The 2001 record of decision led to the most appeals that had been recorded up to that time, and our survey results, reported in detail in the next chapter, indicate that a lack of trust in both the agency and the process remained high.

Although these findings are not unexpected given the complexity of the agency's decision dilemma, they led us to consider decision processes that might both satisfy NEPA requirements and improve outcomes in the case of wicked problems. The US National Research Council (1996), in a study of the challenges associated with promoting effective understanding and management of risk in democratic societies, recommends utilizing learning networks in decision problems involving scientific uncertainty and conflicts in public values.

Learning Networks

The National Research Council report that we draw on here examined problems of risk characterization and management (National Research Council 1996). As we discuss in chapter 2, the presence of risk contributes to several key components of wicked problems. These include (1) uncertainty, driven in part by knowledge gaps and in part by the probabilistic nature of future outcomes; (2) scientific and technical findings, many of which are viewed differently by experts and the public, are poorly understood by the public, or both; and (3) a wide and deep divergence among stakeholders in terms of their risk characterizations, often stemming from fundamental differences in their values and worldviews. Given these overlaps between risk management problems and wicked problems, a decision process designed to clarify and mitigate conflicts concerning risk is also likely to be applicable in the context of wicked problems.

Figure 7.2 offers a graphical illustration of the decision process proposed by the National Research Council (1996, 28). A key point is that the process is both analytic and deliberative. The analytic component is designed to ensure that participants have access to the best scientific and technical information, presented in ways that clarify what is known and what remains uncertain. Moreover the scientific and technical information presented should be relevant to the priorities and concerns of the stakeholders. The report describes these steps as "getting the science right" and "getting the right science" (National Research Council 1996, 6–7). The deliberative component is designed to ensure that the relevant priorities and concerns of the public are recognized, represented, understood, and considered, and that the process is, and is perceived to be, open and transparent. The report describes these steps as "getting the right participation" and "getting the participation right" (National Research Council 1996, 7).

To get the science and participation right, public officials, natural and

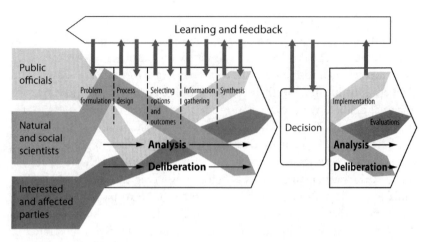

FIGURE 7.2. The learning network concept
Note: Adapted from the National Research Council (1996)

social scientists, and interested and affected parties from the public are included as participating stakeholders. Integrating these three groups directly in the process, and engaging them in ongoing deliberations among themselves, serves to build trust and social capital in ways that do not happen in conventional decision processes (Pretty 2006). For example, scientists are typically engaged in complex decision dilemmas primarily through conducting research, preparing reports, and presenting their findings to agency administrators. Only occasionally do they make presentations to public officials or the public. And very rarely are they drawn deeply into deliberations as stakeholders rather than detached experts. The logic is similar for public officials and for representatives of interested and affected parties from the public. All three groups generally remain separated from each other in the decision process, interacting only in formal settings with clearly distinct roles. These differing roles and priorities may lead to disaffection and miscommunication, and ultimately to adversarial interactions characterized by lack of trust. In such circumstances, decisions are likely to be both harder to reach and less satisfactory.

To break down the barriers of differing roles and priorities, the analytic, deliberative process illustrated in figure 7.2 requires the participants to work together on the challenging tasks of formulating the problem, shaping the process, clarifying preferred outcomes, developing policy alternatives that may help achieve these outcomes, and gathering information necessary for the assessment of the alternatives. In this process, prior to action by the responsible decision maker, the participants synthesize what has emerged from their analysis and deliberations regarding both information and values, and acknowledge remaining limitations, uncertainties, and disagreements. The synthesis aims to satisfy the participants that they have been adequately informed and appropriately heard through the analysis and deliberations included in the decision process. As figure 7.2 indicates, the analytic, deliberative process continues after the formal decision is made, with the focus shifting to implementation and evaluation.

The layout of the main lower section of the figure suggests a process that advances through time in a linear fashion. It begins with the establishment of the participatory process before advancing through problem formulation, information gathering, and so forth. Next, participants formulate a synthesis prior to decision making. The decision is then followed by evaluation of its implementation. The learning and feedback illustrated in the upper part of the figure indicates that there is also an essential circular, iterative component to the process. In principle, the participants' technical knowledge, understanding of divergent values and priorities, awareness of remaining limitations and uncertainties, and trust in the process will grow through mutual learning. Communication in this learning network becomes more fully two way, and participation becomes rich and deep, rather than nominal and superficial. Learning networks have the potential both to produce better decisions (that is, decisions that meet scientific criteria and are satisficing) and to reduce conflicts over the decisions. Uncertainties and disagreements remain, but through the iterative, deliberative, analytic process, they are clarified, made explicit, heard, and discussed.

Our proposed decision approach incorporates this learning network process. The discovery phase, steps 1 through 3 in figure 6.2, includes the problem formulation and process design components of the model recommended by the National Research Council (1996). The deliberation phase, steps 3 through 5 in figure 6.2, then engages key stakeholders in an iterative, deliberative, analytic process to gather information and develop policy alternatives. These steps will foster the emergence of a learning network, as in the National Research Council's proposal. In the aggregation phase, steps 6 and 7 of figure 6.2, participants in the learning network synthesize the results of their mutual learning, leading to the selection of a small set of potentially satisficing policy alternatives. In the evaluation phase, steps 9 and 10 in figure 6.2, our process emulates the postdecision component of the process illustrated in figure 7.2 through engaging participants in monitoring and assessing the outcomes of the implementation. Finally, figure 6.2 makes the iterative nature of the process explicit by indicating that the results of monitoring and evaluation feed back into the steps of modification and adaptation during which, as is essential in the wicked problem context, the outcomes of any decision become the concerns that can be addressed in subsequent rounds of analysis and deliberation.

Incorporating Formal Elicitation and Analysis of Stakeholder Preferences

To further strengthen the modified analytic, deliberative, learning network process illustrated in figure 6.2, we recommend including formal elicitation and analysis of stakeholder preferences. During initial scoping of stakeholder preferences, and even during the iterative, deliberative phase of the process, participants' attitudes and values may not be fully revealed. For example, participants may not have considered their preferences in terms of the inevitable trade-offs inherent in complex decisions. They may not have delved into the nuances and details of various policy alternatives or examined their potential level of satisfaction or dissatisfaction with the range of proposed management strategies and projected ecological and social outcomes. They may not have explicitly articulated their attitudes toward the lead agency or the decision process. Moreover, even if participants are clear about their own attitudes and values, they are unlikely to fully understand the perspectives of others. In contested debates, participants may not express their views in ways that can be heard clearly, or they may stake out negotiating positions that do not reflect their true preferences.

Sophisticated techniques from various disciplines, including marketing, political science, behavioral economics, and others, are available to help clarify the attitudes and values of individuals and groups. Formal elicitation and analysis of stakeholder preferences can generate information that may not emerge from discussion and deliberation. During our research in the Sierra Nevada case, we adapted and pilot tested a set of formal analytic techniques appropriate for use in the context of wicked environmental problems. In the next two chapters, we describe the methods and present the results.

To prepare for that discussion, we suggest here that these techniques can usefully be incorporated in the learning network process illustrated in figure 6.2. In steps 1 through 5 of the discovery and deliberation phases, for example, survey questionnaires, such as the one we describe in chapter 8, can be administered. In steps 3 through 6 of the deliberation and aggregation phase, sorting and ranking exercises like those discussed in chapter 9 can be applied. Data collected through sorting and ranking exercises can support statistical simulations allowing researchers to estimate how participants would respond to combinations of management strategies and projected ecological outcomes beyond those included in the exercises. As discussed earlier, these preference simulations can be integrated with socioeconomic and ecological models to project stakeholder responses to a wide range of scenarios.

Information generated from these analyses and simulations can serve as valuable new input into learning networks to help participants advance their deliberations. The results of simulations, for example, can be particularly useful in steps 4 through 6 of the process illustrated in figure 6.2. As discussed in the next two chapters, our pilot test of the techniques in the Sierra Nevada suggest that formal elicitation and analysis of preference data can reveal promising avenues for negotiation that would otherwise be overlooked.

Application in the Wicked Cases

The decision process depicted in figure 6.2 is designed to improve outcomes in the wicked problem context. It centers on integrating science and values, and providing decision makers with a manageable set of ranked feasible policy alternatives most likely to both meet scientific criteria and be acceptable to a wide range of stakeholders. In this section, we briefly return to the case studies of wicked problems introduced in chapter 3 and consider how components of the approach may be emphasized or adapted to fit the circumstances of each. We examine three of the cases here: the Florida Everglades, Tanzania's Ngorongoro Conservation Area, and Europe's emissions trading scheme. In discussing the pilot test of our analytical techniques in chapters 8 and 9, we review in greater detail how the decision approach may be applied in our fourth example, the Sierra Nevada case.

Each case has characteristics that raise somewhat different obstacles to effective broad-based public participation and the emergence of learning networks of key stakeholders. In South Florida there are large urban areas that depend on the Everglades system for their water supplies but whose populations, unlike those in the local communities of the Sierra Nevada, are not closely connected to the resource base through their vocations, avocations, and traditions. The case of the Ngorongoro Conservation Area in Tanzania presents the added challenge of extreme disparity of political and economic power among key groups. And the EU case is of a different order of magnitude in terms of scale, with more than twenty nations involved.

Nevertheless, we suggest that, despite the obstacles, participatory decision processes incorporating adaptive, deliberative, analytic learning networks can be applied in these cases and can help address the inherent uncertainties and conflicts in these wicked problems.

The Everglades

The Comprehensive Everglades Restoration Plan, authorized by Congress in 2000, marked a significant shift in the approach to public participation in management of the South Florida watershed. The plan incorporates a concerted program of public outreach that significantly exceeds prior limited efforts by the Army Corps of Engineers during implementation of the Central and Southern Florida Project from the 1950s to the 1980s. Nevertheless, there are indications that the process can still be improved

A recent memorandum providing guidance on public participation to staff at the responsible agencies notes that the purpose of public participation is to "(1) provide information on proposed activities to the public; (2) make the public's desires, needs, and concerns known to decision-makers; (3) provide for consultation with the public before decisions are reached; and (4) consider the public's views in reaching decisions" (South Florida Water Management District and US Army Corps of Engineers 2003, 1). Although this list of objectives meets minimum NEPA requirements for participation, it falls short of the high levels of outreach and engagement undertaken by the Forest Service in the Sierra Nevada case, and it clearly does not envision a deliberative process that would encourage full two-way communication with stakeholders and foster the emergence of learning networks.

The document later states that "resolving issues often doesn't require that the entirety of the 'public,' or even a majority of the public, buys into or desires to participate in a decision making process" (South Florida Water Management District and US Army Corps of Engineers 2003, 2). In contrast, our research indicates that progress in addressing wicked problems is unlikely if a majority of the public does not buy into the decision-making process. Decisions reached without this buy-in are highly likely to be subject to appeals and litigation. Clearly in large-scale decision dilemmas such as the Everglades case, agencies cannot be expected to involve the millions of people who may be affected. Nevertheless, best practices described in the literature recommend encouraging the active engagement of stakeholders reflecting as broad a range of public values as possible (National Research Council 1996).

The responsible agencies in the Everglades case, the South Florida Water Management District and the Army Corps of Engineers, could implement the decision process illustrated in figure 6.2 in ways designed to elicit public values on a broad scale across the large geographic area encompassed by the Comprehensive Everglades Restoration Plan. For example, the scoping required under the NEPA process, illustrated in the discovery phase, steps 1 through 3, of figure 6.2, and step 10 of figure 7.1, could include wide outreach throughout the region, both in the urban areas on the east and west coasts of the peninsula and in the agricultural areas of the interior. The survey questionnaire methods we describe in chapter 8 could be employed at public meetings held across South Florida. Information garnered through these processes would be valuable as the more directly involved stakeholders became engaged in the deliberation phase envisioned in steps 3 through 5 of figure 6.2. In this phase both the survey methods described in chapter 8 and the sorting and ranking exercises described in chapter 9 could be conducted. Formal elicitation, analysis, and simulation of participants' preferences would likely improve the satisficing potential of the management alternatives developed in the aggregation phase, steps 6 and 7, of figure 6.2, and steps 11 through 13 of figure 7.1.

In the Everglades case, rapidly shifting political alliances and severe financial constraints in the current economic downturn have exacerbated the issues of large geographic and demographic scales, scientific uncertainty, and divergent public values that commonly contribute to wicked problems. Nevertheless more effective participatory processes built around learning networks have the potential to strengthen resolve and improve outcomes as participants grapple with the decision dilemma.

The Ngorongoro Conservation Area in Tanzania

Wicked problems stem in part from profound differences in the values, beliefs, and worldviews of stakeholder groups. Commonly, the various interested parties also differ significantly in the levels of political power and financial resources they can draw on as they attempt to influence management choices. In developed democratic societies, social norms and legal frameworks typically support principles of equal access. Although these safeguards do not eliminate differences in influence, they do buffer the effects of inequities in money and power.

In the Ngorongoro case, however, the disparities are far more extreme. International tourists who visit the region, and the tourism industry that provides their services, contribute approximately a billion dollars a year to the Tanzanian economy, about 15 percent of the nation's gross domestic product. In contrast, the majority of Maasai people who live in and around the Ngorongoro Conservation Area are poor, illiterate, and politically marginalized. As discussed in chapter 3, 60 percent of the Maasai in Ngorongoro live below the poverty line, and 40 percent are categorized very poor or destitute (McCabe 2003).

These severe socioeconomic inequalities create obstacles to fair and equitable participatory processes. The concept of a learning network assumes equal access and influence among participants. Each group is assumed to bring strengths to the process, and each group's underlying beliefs are respected. A learning network, as we have adopted the concept, brings together scientists, public officials, and interested and affected parties from the general public (National Research Council 1996). While scientists bring technical knowledge that the other participants may not have, the learning network process assumes that the other participants can learn the essential points of the relevant science. Similarly, the approach assumes that the technical experts are open to understanding the differing value systems and perspectives of the public stakeholders and can benefit from local knowledge. The participants can deliberate together on equal footing. In the Ngorongoro case, this sense of equal social standing and equal access to information and influence is substantially more difficult to achieve.

To help address these challenges, we suggest that the management agency should make a particularly strong effort early in the process to elicit the values and preferences of the Maasai pastoralists. Rural development workers and agricultural extension experts, among other professionals, have developed effective, culturally sensitive ways to gather information from and pass information to marginalized, poorly educated residents of rural areas in poor countries. In Tanzania, several national ministries, including those focusing on community development, agriculture, and health, along with the Ngorongoro Conservation Area Authority, have staff with these skills.

In adopting the decision approach illustrated in figure 6.2, the management agency could train staff members skilled in community outreach to conduct the types of survey questionnaires and sorting and ranking exercises that we recommend. In this way, Maasai concerns could be introduced systematically into the participatory, deliberative process. Moreover the outreach workers could encourage and support the participation of Maasai representatives in the ongoing learning network that should emerge from the deliberations.

As in the other wicked problem cases, the process we recommend will not serve as a panacea in the Ngorongoro decision dilemma. But it holds promise for promoting more equitable and effective participatory processes in circumstances such as those in rural Tanzania, where deep social and political inequities limit the efficacy of conventional approaches to public engagement.

Europe's Emissions Trading System

The contested multinational effort to implement emissions reduction policy in the European Union presents a different sort of higher-order challenge to effective participatory processes. As we have noted, larger scales—geographic, demographic, political, temporal—increase the likelihood that wickedness will emerge in public management decision dilemmas.

The decision problems in the Sierra Nevada, South Florida, and Tanzanian cases all encompass extensive and diverse ecological landscapes, long time horizons, large-scale and dynamic socioeconomic and demographic conditions, and political and budgetary influences from the local, subnational, and national levels. And in all three cases, problems related to scale have contributed to unsatisfactory outcomes. In the European Union's effort to mitigate greenhouse gas emissions, these problems are magnified and compounded.

In the aggregate, the European Union, with approximately 500 million people and an annual combined gross domestic product of approximately $14.5 trillion, has a population larger than any country except China and India and a combined economy that is now the largest in the world. Moreover, politically the European Union is a federation of twenty-seven independent nations with diverse cultures, histories, and governance structures, and the union's central institutions, which lack deep roots and well-established legitimacy, are regarded skeptically by many citizens.

To address these potentially overwhelming problems of scale, our recommended process, illustrated in figure 6.2, might best be implemented in stages. That is, NEPA-type participatory decision processes, with the recommended enhancements of learning networks informed by formal elicitation and analysis of stakeholder preferences, could be implemented at the subnational and then national level in each member country. The process could then be instituted at the central EU level. At each stage, the process should reflect the full richness of the deliberations at prior levels, thus ultimately capturing the divergence of views both within and across member countries. At each level, select key stakeholders from the learning networks that emerged at prior stages could serve as representatives of their regions in the learning networks developed at the subsequent stage. At each level, the process of formal elicitation and analysis of stakeholder preferences could be replicated to generate useful new input and explore opportunities for negotiation and progress that might otherwise be missed.

Clearly, implementing the process in a decision dilemma of this scale would be a challenging endeavor with many potential pitfalls. Unforeseen complications and sources of frustration are certain to emerge. Yet large scales and political complexity are already fundamental characteristics affecting all aspects of decision making in the union. EU policy makers constantly struggle with problems of trust, legitimacy, authority, engagement, and effectiveness. Implementation of the decision approach we recommend would be subject to the same pressures that all decision processes face in this exceptionally demanding management environment. Yet its key elements—designed to incorporate divergent views and adapt to uncertainty and conflict—suggest that it may provide improvements over current practices.

Conclusions

In this chapter we have introduced a decision process that integrates learning networks within the NEPA process and adds the novel component of formal elicitation and analysis of stakeholder preferences. This approach is designed to address the conflicts and uncertainties that often derail efforts to achieve implementable decisions in the context of wicked problems. Rather than raising expectations for the identification of an optimal management choice that will resolve the decision dilemma, this process encourages stakeholders to accept the more modest but perhaps more realistic goal of iteratively identifying feasible management alternatives that lead incrementally and adaptively in the direction of satisficing outcomes.

In meeting NEPA requirements, the proposed process is applicable in the practical statutory and regulatory decision-making environments many public managers face, both in the United States and elsewhere. In incorporating the learning network concept, it adopts best practices recommended by scholars from a range of disciplines who have studied obstacles to effective decision making in complex cases. As suggested in our discussion of its possible application in the wicked problem case studies, the process may be flexible enough to be adapted usefully to a range of challenging circumstances.

Chapter 8

The Sierra Nevada Example: Survey of Stakeholders

We turn now to practical aspects of the elicitation and analysis of stakeholder preferences and the application of the results to the *learning network* approach. We suggest that the techniques proposed in this chapter and the next can provide valuable insights into stakeholder preferences and their consequences that might otherwise be overlooked. The linking of preferences to potential consequences may help clarify the role of values in public participatory processes.

For this detailed discussion of preference elicitation and analysis, we focus on the Sierra Nevada Forest Plan Amendment process, in which the Forest Service has worked to develop a broadly acceptable plan for managing the national forests in the Sierra Nevada region of California. Figure 8.1 shows the location of all the national forests that fall within the region. The map also identifies those forests that were part of the Sierra Nevada Forest Plan Amendment process. These forests cover more than eleven million acres, and about four million people live in their immediate vicinity. The map also shows county borders. The geographic scale, demographic diversity, and number of political jurisdictions in the Sierra Nevada region contribute to the management challenge.

During our work for the Forest Service related to the Sierra Nevada Forest Plan Amendment decision dilemma, we were able to collect data

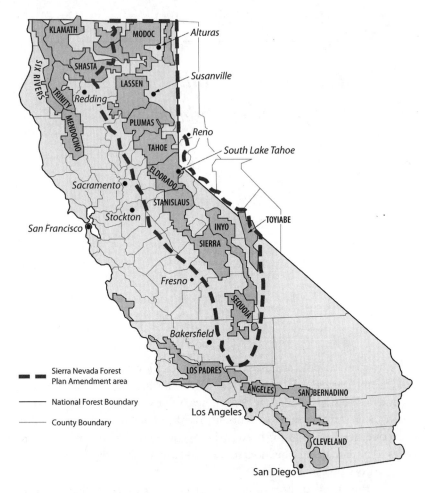

FIGURE 8.1. The national forests of the Sierra Nevada region
Note: Adapted from USDA Forest Service Pacific Southwest Region maps

from active participants in the process. In collecting data, we used two instruments: a survey questionnaire and a card-sort exercise. In this chapter we focus on the survey questionnaire, particularly on how we obtained the data and what we learned about the stakeholders from their responses. In the next chapter we focus on the results of the card-sort exercise.

In analyzing data collected through application of these two instruments, we were able to take the first steps in pilot testing our proposed techniques for enhancing the effectiveness of learning networks in the context of wicked problems. We note at the outset that we had a relatively small sample of participants in the forest planning process and that our respondents were not randomly selected. Thus we do not claim that the specific results from our analyses are generalizable. Instead we view our research as a trial of what we believe is an innovative approach to the elicitation and analysis of stakeholder preferences that may make a useful contribution to improving outcomes when learning networks are established.

Contextual Background

Public values toward the national forests have changed dramatically since their creation out of the public domain beginning in 1891. Born out of the Progressive Era of Gifford Pinchot and Teddy Roosevelt (Pinchot 1947; Nelson 1999), the national forests have historically been managed to serve utilitarian purposes, especially in the western United States, to meet the people's needs for wood, water, forage, and other natural resources important for economic development. For the majority of the national forests' existence, this land management philosophy received general support from Congress and the public. Today, though, a wider range of public values has emerged. Some, especially in local communities and industries that depend on the national forests for livelihoods and inputs, continue to place a priority on conventional economic uses of the forests. Others, such as recreationalists seeking broader access to public lands, may favor the construction of roads and trails for motorized vehicles. Still others, often identified as environmentalists, may oppose any extractive activities in the forests, especially timber harvesting but also livestock grazing and mining, and also oppose recreational activities, such as off-road use of motor vehicles, which may produce environmental degradation or interfere with the enjoyment of the forests' spiritual and esthetic values.

Changing demographics in rural Sierra Nevada communities mean that these conflicting societal values play out in local politics and planning efforts, such as those facing the Forest Service. Old-line community members, who obtain their livelihoods from commercial uses of the national forests, increasingly come into conflict with newer residents, who often move to the communities to escape hectic urban lifestyles, and who thus have different expectations for forest management. Complicating matters further, diverse national interest groups favoring widely divergent management approaches and objectives commonly apply political pressure to promote adoption of their policy preferences. Each of these disparate groups views forest outcomes from its own perspective and expects the forests to be managed in accordance with its values. If these groups perceive their rights or values to be in jeopardy, they often turn to litigation or other forms of political or administrative pressure to influence outcomes.

The Management Dilemma

In the context of the national forests, the regional forester is the ultimate decision maker responsible for the day-to-day management. In this role, as illustrated in figure 8.2, the regional forester must act in a dynamic, complex decision environment characterized by ecological, political, administrative, and stochastic uncertainties.

In characterizing this complex and dynamic decision environment, we find three main drivers influencing forest ecosystems over time:

- external human factors, associated with both neighboring communities and the broader interested public;
- natural events and processes largely beyond the agency's control— including lightning strikes, droughts, plant community succession, and long-term changes in climate; and
- Forest Service management strategies and practices.

These effects are then translated into outcomes, such as wildfire acres burned, old-growth habitat gained or lost, and timber volume produced. In turn, these outcomes, both observed and projected, influence stakeholder acceptance of, and reaction to, Forest Service management activities. But stakeholders also evaluate agency decisions on factors other than ecological outcomes. Whether because of scientific uncertainty, lack of trust in the agency, or some other reason, stakeholders also scrutinize and evaluate the Forest Service based on the processes it uses in reaching its management decisions.

In addition to holding diverse preferences regarding processes and outcomes, stakeholders differ in their understanding of—and aversion to or tolerance for—the range of risks and uncertainties associated with forest management. Stakeholders, acting on their preferences and levels of risk aversion or tolerance, then influence the Forest Service's management strategies and practices through various mechanisms, including participating in public meetings and other opportunities for public comment, bringing political influence to bear, and initiating formal appeals and legal challenges.

In principle, the development of a learning network of stakeholders has the potential to moderate these conflicts. A learning network aims to facili-

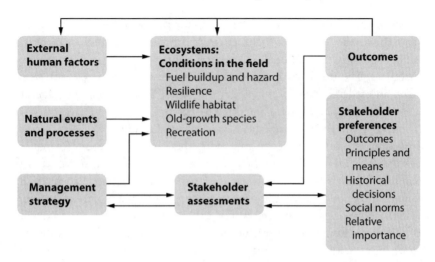

FIGURE 8.2. The Forest Service decision problem

tate positive interactions among stakeholders (National Research Council 1996). It includes processes designed to provide participants in the network with pertinent information about what is known and what remains uncertain regarding scientific, social, economic, political, and administrative states and trends. It then offers stakeholders opportunities to express their preferences in the context of the information available and to consider the implications of their policy preferences for likely forest outcomes. As the learning network matures, trust should build. Iterative cycles involving the exchange of information and the discussion of preferred outcomes may promote the emergence of an analytic, deliberative process that improves the chances for a broadly accepted, implementable decision. We suggest that this process can be improved by systematically providing participants in the learning network with deeper information about stakeholder preferences derived from formal elicitation and analysis of attitudinal data. Following the process articulated in chapter 6, we use our study of the Sierra Nevada case as an example of how such systematic process of data collection and analysis might work.

Data Collection

In 2003, we held three workshops to elicit attitudes and preferences related to the Sierra Nevada Forest Plan Amendment process. Participants included individuals who had been actively engaged over an extended period, either as interested members of the public or as employees of relevant government agencies. Two of the workshops were open to stakeholders from the public. Attendees included, for example, representatives from the timber industry, recreationalists, environmentalists, and local community members. The third workshop was held for employees of relevant local, state, and federal agencies, including the Forest Service. The same agenda and content were covered at the three workshops.

To prepare for the workshops, we reviewed a number of documents. In addition to documents accompanying the regional forester's formal record of decision in 2001 and the public's response to that decision, we also studied the records of the various efforts the Forest Service had made to engage the public during the process leading up to the record of decision (USDA Forest Service 2001a, 2001b, 2003). We also had access to details of various scientific studies, including models assessing and forecasting decades into the future the extent of old-growth owl habitat, the incidence of wildfires, patterns of demographic and economic development, and other relevant parameters. In addition, we were able to interview regional foresters and other agency officials who had faced similar forest management decisions in the past in the Sierra Nevada and in other regions.

In inviting participation from members of the public, we advertized the workshops on the Forest Service's Sierra Nevada Forest Plan Amendment website and in the newspapers of affected communities. We also directly invited 55 individuals whom the Forest Service identified as having been actively involved with the development of the environmental impact statement and record of decision in the Sierra Nevada Forest Plan Amendment case. Forest Service officials indicated to us that at the start of the Sierra Nevada planning process in the mid-1990s there had been as many as 250 actively engaged individuals from the public. Not surprisingly, that number dwindled as the process dragged on. So the 55 invitees represented those who had been most concerned and committed over the long term.

We held the third workshop with stakeholders from the relevant government agencies. We scheduled this workshop in conjunction with a meeting of the regional interagency team, which included representatives of federal and state land management and regulatory agencies, wildlife and forest scientists, and key Forest Service personnel. Participants from the Forest Service included the regional forester, assistant regional foresters, forest supervisors, and other senior staff responsible for forest planning and management. We invited 155 members of this group to the workshop. This number included the 12 local heads of federal and state agencies constituting the executive committee of the interagency team.

Thus, in total we sent out 210 invitations to individuals, 55 to stakeholders from the public and 155 to stakeholders from relevant government agencies. In all, 77 people participated in one of the three workshops. Of these 77 people, 36 attended the public workshops and the remaining 41 attended the workshop held for the various government agencies. At all three workshops, we provided an overview of the Forest Service's decision dilemma, including an introduction to the concepts of wicked problems, risk and uncertainty, and participatory processes. We then asked participants to complete the survey questionnaire and card-sort exercise to help us collect attitudinal data. These activities were designed to help us identify attitudes toward, and preferences for, such factors as decision-making processes, approaches to risks and trade-offs, general environmental management strategies, specific management priorities for the Sierra Nevada national forests, and the performance of the Forest Service as a management institution. Two participants chose not to participate in these activities, so we ultimately collected data from 75 respondents.

As noted earlier, we worked in this case with a relatively small sample of respondents that was not randomly selected. We cannot say, based on our research, how other stakeholders who were not present may have responded to our data collection exercises. Nonetheless, all of our respondents were either directly involved in the Sierra Nevada Forest Plan Amendment process or were citizens and government officials with long-term engagement in forest planning in the region. As such, their participation gave us the opportunity to pilot test our techniques, including the innovative analytic methods we describe in more detail in the next chapter. As we discuss in this and the next chapter, these methods have the potential to provide new and valuable input into learning networks established to address wicked environmental problems.

Findings from the Survey Questionnaire

The questionnaire included four parts. In the first part, we presented respondents with a series of forty-eight statements and asked them to indicate whether they agreed or disagreed with each statement. Response options were on a five-point scale: strongly disagree, disagree, neither agree nor disagree, agree, strongly agree. Tables 8.1 through 8.5 report aggregated responses to selected statements from this section of the questionnaire. While we discuss the responses to all the questions, we present only those where there was substantial variation in the responses among the different groups.

In the second part of the questionnaire, we presented respondents with a list of twenty potential forest management priorities and asked them to rank these priorities on a scale of one to ten based on their own preferences (respondents could give the same ranking to multiple priorities). In the third part, we presented two open-ended questions asking respondents to identify the greatest risk faced by the Forest Service and by the public in the Sierra Nevada management dilemma. We also discuss the results of the second and third parts of the questionnaire.

Finally, we asked respondents to provide some information about themselves, including the length and intensity of their involvement with the Sierra Nevada planning process and their relevant organizational affiliation, if any. In analyzing responses to the questionnaire, we grouped participants into three categories based on their self-identified affiliations:

- members of the public (47 percent of the respondents belonged to this group), including representatives of private-sector or business-related organizations; members of environmental or other nongovernmental organizations; individual concerned citizens; and all others not elsewhere classified;
- employees of government agencies other than the Forest Service (21 percent), including municipal or county government employees, state government employees, and federal government employees in agencies other than the Forest Service; and
- Forest Service employees (32 percent).

Potential for Consensus and General Level of Satisfaction and Trust

In examining responses to the statements in the first section of the questionnaire, we found that high percentages of all groups were generally pessimistic about the potential for finding a consensus agreement for managing the Sierra Nevada. When asked to evaluate the statement, "A consensus agreement is possible that would satisfy all participants concerned about management of the Sierra Nevada forests," 79 percent to 83 percent of respondents in all three groups disagreed or strongly disagreed with the statement. Moreover, as table 8.1 indicates, nearly half of all participants did not feel that their most important concerns had been adequately incorporated in the process. The sense that their concerns had not been adequately addressed was also true for a substantial minority of Forest Service employees.

What is less clear is whether the dissatisfaction with the process was due to disagreement with the outcome or frustration with the process

TABLE 8.1. Summary of responses to statements regarding the Sierra Nevada Forest Plan Amendment (SNFPA) decision process

Statement	Public		Other government employees		Forest Service employees	
	SD/D	A/SA	SD/D	A/SA	SD/D	A/SA
The issues and concerns that I believe are most important are adequately incorporated in the SNFPA decision process.	60%	20%	43%	50%	35%	43%
The SNFPA decision process affords adequate opportunity for public involvement and deliberation in determining the final management goals and priorities.	46%	30%	13%	53%	50%	41%
The SNFPA decision process has resulted in increased agreement among most parties on what the most important issues and goals should be.	68%	8%	50%	36%	50%	27%
My trust in the US Forest Service and its management of the Sierra Nevada has increased as a result of the SNFPA decision process.	62%	19%	27%	47%	36%	36%
The SNFPA decision process is helpful in educating and informing the public on key decision issues.	35%	41%	13%	60%	18%	59%
On balance, I believe that my participation in the SNFPA decision process has improved the quality of the final decision.	28%	53%	7%	73%	9%	77%

Note: SD = Strongly disagree; D = Disagree; A = Agree; SA = Strongly agree. The "neither agree nor disagree" responses are not shown.

itself. The findings summarized in table 8.1 reveal that half the Forest Service employees and a plurality of the public disagreed that the process had afforded adequate opportunity for public involvement. We further refined the analysis by cross tabulating the responses to gain insights into how respondents who had a specific response to a question also felt about other questions. For instance, among those who felt their most important concerns were not adequately incorporated in the process, 65 percent also felt that the opportunity for public involvement was inadequate. Clearly, these participants could be rejecting a process that yielded what was for them an unacceptable result. Yet even among those who felt their views were adequately incorporated in the decision process, 50 percent did not agree with the statement that opportunities for public involvement and deliberation were adequate. It would certainly not be a surprise to find that those who objected to the decision outcome might have found fault with the decision process. At least for a strong minority of participants, however, the Forest Plan amendment process clearly needed further improvement.

A majority of public participants and a substantial minority of Forest Service employees disagreed with the statement that their trust in the Forest Service had increased because of the process. Again, this erosion of trust seemed strongest among those who did not feel their important concerns were adequately incorporated in the process (71 percent). But even among those who were either neutral or agreed that their concerns were incorporated, 58 percent did not agree with the statement that their trust in the Forest Service had increased.

Regardless of their attitudes toward the decision outcomes, however, the results in table 8.1 also indicate that a plurality or majority of members of all three groups felt that their personal contributions had made a difference and that the process had been valuable in educating the public. These results provide modest support for some optimism that progress may be possible.

Forest Service Capacity

Table 8.2 reports on participants' attitudes toward the Forest Service and its capacity to manage the Sierra Nevada national forests, as revealed in their responses to statements related to this issue. Generally, most participants felt that the Forest Service had or could obtain the technical skills necessary to manage the forests. However, the majority of respondents outside the

Forest Service did not agree that the Forest Service had a good fire management record. Moreover, there was disagreement about whether the Forest Service could be trusted to protect and restore owl habitat. Thus, while most participants saw the agency as competent (or at least potentially competent), outsiders had concerns about both the record of accomplishments to date and the Forest Service's priorities for the future.

While keeping in mind that this was a relatively small, partially self-selected sample from a population of people engaged with forest related issues, these responses suggest a certain lack of faith in the Forest Service. However, when viewed in a larger context, these opinions are not at odds

TABLE 8.2. Summary of responses to statements regarding the capacity of the Forest Service

Statement	Public		Other government employees		Forest Service employees	
	SD/D	A/SA	SD/D	A/SA	SD/D	A/SA
The Forest Service has, or can develop, the skills and information necessary for effective medium- to long-range ecological risk management in the Sierra Nevada forests.	8%	67%	14%	71%	22%	70%
The Forest Service has a good fire management record in the Sierra Nevada.	54%	19%	57%	14%	13%	70%
Forest Service personnel can be trusted to protect and restore essential habitat for the California spotted owl and other old forest species.	38%	32%	14%	57%	17%	74%
Unexpected outcomes from management actions are the result of agency failings.	59%	11%	57%	7%	87%	4%

Note: SD = Strongly disagree; D = Disagree; A = Agree; SA = Strongly agree. The "neither agree nor disagree" responses are not shown.

with the public's general lack of faith in government and in public institutions or in the ability of such organizations to protect individuals from harm. One purpose of a learning network's extended engagement activity is to build trust, both in the Forest Service's ability to act in a just and socially responsible fashion and in the public's ability to provide useful and well-informed input.

The Decision Process and the Value of Public Participation

Views on the appropriate decision processes to follow in managing the Sierra Nevada national forests were ambivalent. As shown in table 8.3, most participants did not feel, based on their responses to statements in the questionnaire, that the public had the expertise necessary to manage the forests. At the same time, there was substantial disagreement over the role of experts. Over half of all participants disagreed with the statement that expert plans were more feasible and balanced than those developed by local participants. Even 65 percent of Forest Service employees took that position. In addition, a majority of all participant groups disagreed with the statement that management decisions should be guided solely by science and expert opinion.

With regard to the role of public values and participatory processes, while most groups agreed that these were important, support was strongest among Forest Service employees. On balance, it appeared that most participants—while recognizing that citizens must be educated and informed if they are to participate effectively—remained skeptical of claimed expertise and continued to value local stakeholder views and engagement.

That view, if borne out in the general public, further emphasizes the eroding status of experts and expert knowledge in decision making. The success of the public participation process we are proposing depends on the various stakeholders appreciating multiple perspectives, accepting the inevitability of trade-offs, and being willing to make concessions.

Trade-offs

In responding to statements in the questionnaire related to trade-offs, participants were asked to consider detailed numerical options for socioeconomic and ecological outcomes. Overall as indicated in table 8.4, the responses to questions regarding trade-offs inherent in management de-

TABLE 8.3. Summary of responses to statements regarding public and expert participation in decision processes

Statement	Public		Other government employees		Forest Service employees	
	SD/D	A/SA	SD/D	A/SA	SD/D	A/SA
The general public lacks the specialized knowledge necessary to guide management decisions in the Sierra Nevada.	30%	59%	13%	73%	27%	55%
Forest management plans developed by experts are generally more feasible and balanced than plans developed by local participants.	46%	38%	40%	53%	65%	22%
Management decisions in the Sierra Nevada should be guided solely by science and expert opinion.	72%	8%	67%	13%	86%	5%
Broadly held public values should guide management decisions in the Sierra Nevada.	27%	51%	33%	20%	23%	64%
Democratic, participatory processes generally lead to better forest management decisions than processes dominated by experts.	24%	49%	20%	67%	0%	83%

Note: SD = Strongly disagree; D = Disagree; A = Agree; SA = Strongly agree. The "neither agree nor disagree" responses are not shown.

cisions indicated a sophisticated understanding of underlying issues. Participants both acknowledged and demonstrated the willingness to confront trade-offs between, for example, the likelihood of catastrophic fires and damage to old-growth forest habitat.

TABLE 8.4. Summary of responses to statements regarding trade-offs

Statement	Public		Other government employees		Forest Service employees	
	SD/D	A/SA	SD/D	A/SA	SD/D	A/SA
A 3 percent short-term reduction in old forest habitat acreage is acceptable if there is a good chance that over the long term at least a 10 percent gain in habitat acreage will result.	27%	61%	0%	93%	14%	86%
A 3 percent short-term reduction in old forest habitat acreage is acceptable if there is a good chance that long-term economic benefits to adjacent communities will result.	36%	48%	14%	57%	14%	76%
A 3 percent short-term reduction in old forest habitat acreage is acceptable if there is a good chance that over the long term safety benefits to adjacent communities from reduced fire hazard will result.	33%	61%	0%	79%	10%	86%
Providing fuels treatments on 1.5 to 2 percent of the forest each year (roughly 150,000 acres) is acceptable if there is a good chance the average number of acres burned will be reduced by 5 to 10 percent per year.	22%	69%	15%	77%	5%	95%

Note: SD = Strongly disagree; D = Disagree; A = Agree; SA = Strongly agree. The "neither agree nor disagree" responses are not shown.

TABLE 8.4. *(continued)*

Statement	Public		Other government employees		Forest Service employees	
	SD/D	A/SA	SD/D	A/SA	SD/D	A/SA
A 3 percent short-term decline in spotted owl nesting habitat is acceptable if there is a good chance the average number of acres burned will be reduced by 5 to 10 percent per year.	31%	56%	23%	54%	13%	74%
A 3 percent short-term decline in spotted owl nesting habitat is acceptable if there is a good chance the average acres in lethal or stand-replacing fires will be reduced by 10 to 30 percent per year.	25%	63%	8%	77%	9%	87%
The creation of small openings or gaps in the forest canopy is acceptable if there is a good chance that the long-term effects on forest regeneration and health are positive.	15%	82%	7%	93%	13%	87%

Table 8.4 indicates that strong majorities of participants from all three groups were willing to trade off short-term losses in old-growth forest habitat for a likelihood of long-term habitat gains. They were less likely to make this trade-off for potential economic benefits. Majorities in the three groups, however, favored all trade-offs that were presented as having the potential to reduce the risk of wildfire. These results are at odds with the general impression we noted among Forest Service administrators that some members of the public, particularly environmentalists, may be unwilling to make such trade-offs. Participants from all groups appeared to accept that trade-offs are unavoidable and appeared willing to consider

recommendations that included undesirable outcomes resulting from these unavoidable trade-offs, particularly if they were associated with the reduced threat of dangerous wildfires.

Participants also demonstrated a slight preference for the location of fuels treatment—efforts to thin vegetation in the forests to reduce combustible material available as fuel for wildfires. Respondents generally preferred that fuels treatment occur in the so-called wildland-urban intermix—zones within 1.5 miles of residences and other improvements—rather than elsewhere in the forest. They also believed that some costs of managing the forests should be recoverable through harvesting of trees and other commercial uses of natural resources.

Management Philosophy

As indicated in table 8.5, participants' responses to statements on management philosophy indicate they generally agreed that, given underlying uncertainties, some form of adaptive management may be the best approach to managing the forests. They were willing to accept some adverse outcomes resulting from experimentation with different policy strategies in order to learn more about the consequences of implementing such strategies.

Responses to the first and third statements listed in table 8.5 show further that workshop participants did not favor strict application of the precautionary principle. In fact, they appeared willing to tolerate some risk of harm in order to learn from experimentation. These results counter the general preconception among Forest Service managers, discussed earlier, that the public generally favors precautionary approaches and distrusts adaptive management.

Management Priorities

In the second section of the questionnaire, we asked participants to rank management priorities for the Sierra Nevada national forests. Participants' top priorities were as follows (the number in parentheses represents the percentage of respondents ranking this option either first or second in importance):

- complying with all environmental and legal requirements (64%);
- following a decision process that is open and fair (63%);

TABLE 8.5. Summary of responses to statements regarding management philosophy

Statement	Public		Other government employees		Forest Service employees	
	SD/D	A/SA	SD/D	A/SA	SD/D	A/SA
When outcomes of management decisions are uncertain, the safest course is to take no action.	88%	9%	85%	8%	96%	4%
When outcomes of management decisions are uncertain, adaptive management is the most responsible approach.	12%	70%	0%	77%	13%	70%
If a management or research "experiment" may damage some habitat for an old-growth, forest-dependent species, then the experiment should not be allowed.	61%	27%	85%	0%	91%	4%

Note: SD = Strongly disagree; D = Disagree; A = Agree; SA = Strongly agree. The "neither agree nor disagree" responses are not shown.

- avoiding catastrophic fire losses in communities (63%);
- protecting threatened and endangered species (59%);
- promoting good air quality (55%);
- enhancing healthy and abundant old forest habitat (54%); and
- avoiding catastrophic fire losses in old forests (52%).

These priorities foreshadow some of the results of the card-sort preference elicitation exercise discussed in the next chapter. In particular, and perhaps in contrast to the Forest Service administrators' preconceptions, even the most ardent environmentalists participating in our workshops were not entirely averse to some timber harvesting. People recognized that a decades-long policy of aggressive fire fighting had been counterproductive in that it had led to the buildup of enough combustible material in the forests to increase the probability of unmanageable and catastrophically damaging wildfires.

Even in the case of a controversial issue like harvesting trees, the notion that there were groups that were adamantly and inflexibly opposed to it turned out not to be the case, at least among our respondents. And we note that our respondents included outspoken environmental advocates. Participants from across the spectrum recognized that a total ban on harvesting was not feasible. The discussion in the workshops focused instead on where timbering should be permitted and what limit should be set on the diameter of trees to be cut. Individuals generally supportive of logging favored allowing timber harvesting in more areas and allowing larger diameter trees to be cut, while those more skeptical of logging favored stricter limits on where harvesting could occur and smaller diameter limits on which trees could be cut.

Conclusions

In summary, according to the respondents, the participatory decision process implemented by the Forest Service had not increased trust in the agency, increased consensus among stakeholders, or provided adequate opportunity for public involvement and deliberation. Moreover, although the respondents acknowledged that there were multiple risks and indicated that they valued them differently, they appeared to understand and be willing to accept trade-offs among the competing risks. Also, in contrast to the preconceptions of Forest Service administrators, we found that participants generally recognized that strict application of the precautionary principle was impractical. In fact, respondents across all groups accepted that some form of adaptive management involving experimentation and learning was a preferred approach to forest management.

Finally, we also learned during the workshops that even within the three categories of respondents discussed in this chapter—the public, government employees with agencies other than the Forest Service, and Forest Service employees—there was a significant diversity of opinion. We explore this point further in the next chapter, where we report the results of our analysis of data from the card-sort exercise.

Chapter 9

The Sierra Nevada Example: Elicitation and Analysis of Preferences

The survey of the workshop participants described in chapter 8 offered a broad overview of the multiple stakeholder perspectives regarding the management of the national forests in the Sierra Nevada region. Even among groups that seem superficially homogeneous, such as individuals who work for the Forest Service, we found a diversity of opinions about how best to manage the forests.

Given the litigious history of efforts to develop a management plan for these forests, one of our preconceptions coming into the process was that the stakeholders would typically have entrenched positions with little room for negotiation and compromise. We learned, however, that while participants often did have these differing positions, they also generally recognized that the status quo was not sustainable and that there had to be some flexibility and new thinking. As discussed in the previous chapter, responses to the survey questionnaire gave us insights into these opinions, including who held them, where there was common ground, and where there were strong differences. The results of the questionnaire gave us a better understanding of the opinions of the various groups of respondents, but we still did not have a clear understanding of individual priorities and attitudes toward potential trade-offs.

We followed up the survey questionnaire with a preference elicitation exercise that explored in finer detail the attitudes of individual respondents toward various combinations of management practices and potential outcomes. This exercise allowed us to determine which sets of trade-offs individuals might consider and which ones they would find unacceptable. Figure 9.1 provides a road map of the process we followed for obtaining the individual preferences and analyzing the results. We used two separate statistical techniques for analyzing the data, and we supplemented each analysis with a simulation. We envision that the iterative, adaptive, deliberative process (see fig. 6.2) to be applied in the case of wicked problems will be informed by continuing iterations of the process illustrated in figure 9.1. In addition to highlighting areas of agreement and disagreement among the stakeholders, this process of preference elicitation and analysis serves at least two purposes: it makes the preferences of the stakeholders explicit, both for the stakeholders themselves and for other participants, and it clarifies the potential consequences of policies that might result from acting on those preferences.

Decision Criteria

In preparation for the workshops we met with Forest Service officials, attended scientific meetings, reviewed the literature, and relied upon our own knowledge of the issues to obtain a manageable set of variables that would serve as the basis of our efforts to elicit stakeholder preferences. Our objective was to identify attributes of management alternatives and projected outcomes that stakeholders cared about in terms of the choices the Forest Service could make in managing the forests.

In order to reduce the risk of wildfire, for example, the Forest Service may attempt to reduce the amount of fuel available for potential forest fires by removing understory vegetation and/or thinning the forests. Yet given that the Sierra Nevada national forests cover over eleven million acres, with over seven million acres of the land forested, the agency cannot practically aim to eliminate dangerous fuel loads across the entire management area. Some areas will be treated while others will not. In selecting forested areas for treatment, the agency has to decide how to allocate scarce resources among areas near human settlements, areas with endangered species and habitat, and other general forested areas. The areas near human settlements are known as *wildland-urban intermix* (WUI) areas, which in turn are subdivided into so-called defense zones directly adjacent to developed areas and threat zones that serve as buffers between the defense zones and the general forested areas.

The Sierra Nevada Example: Elicitation and Analysis of Preferences 169

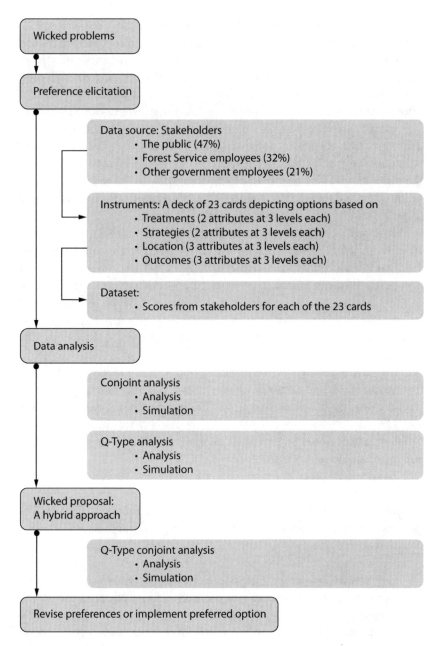

FIGURE 9.1. Analytical road map

Moreover, both of the two primary means of reducing excess fuel—prescribed fires (fires intentionally set and controlled by the agency) and mechanical treatment (use of timber harvesting machinery and techniques)—are problematic and controversial. Prescribed fires can burn out of control, and even when properly contained these fires can produce significant air pollution problems from smoke and ash. Mechanical treatment can create unsightly damage in the forests and can exacerbate distrust of the agency, as some stakeholders see it as a way to circumvent bans on timber harvesting, particularly in old-growth habitat. On the other hand, both treatment practices can provide forest management benefits beyond fire control, for example by opening gaps or holes in the forest canopy to promote regeneration.

Costs are a further complicating factor. Fuel treatment is expensive. In principle, mechanical treatment that included harvesting of marketable trees could generate revenue to help defray the costs. Yet timber harvesting in areas linked to threatened and endangered species, such as spotted owls and their old-growth habitat, is severely constrained by law. An additional challenge is that there is substantial scientific uncertainty in predicting the outcomes of fuel treatment policies over the short and long term.

Thus the Forest Service has to choose both where and how to engage in fuel treatment. These decisions involve unavoidable trade-offs—trade-offs that deeply engage, often in sharply differing directions, the preferences and values of a wide range of stakeholders. As is characteristic of wicked problems, the regional forester in such cases is forced to act in a contentious and uncertain decision environment.

At our initial meetings with participants, we had begun to get a sense of the attributes of the choices that the regional forester faced. Hence for the preference elicitation exercise, we settled on the following four dimensions as the basis for our study of the fuel treatment dilemma in the Sierra Nevada context:

- *treatments*—whether to use prescribed fires and (or) mechanical removal, and the percent of forested land to undergo each treatment;
- *strategies*—regarding the amount of timber and wood salvage offered for sale, and the extent of forest gaps to be created;
- *locations*—where the treatments should occur, whether in the defense and threat zones of the wildland-urban intermix areas or elsewhere; and

TABLE 9.1. Attribute definitions and descriptions

	Attribute description	Attribute name	Levels
Treatments	**Mechanical—percentage of forested land (1st decade):** The total acres treated over the next ten years using mechanical fuels treatment, as a percentage of total forested lands in the eleven Sierra Nevada national forests (7,372,257 acres).	Mechanical	1% 5% 10%
	Prescribed fire—percentage of forested land (1st decade): The total acres treated over the next ten years using prescribed fires as the fuels treatment, as a percentage of total forested lands in the eleven Sierra Nevada national forests (7,372,257 acres).	Prescribed fire	5% 8% 12%
Strategies	**Change in timber and salvage offered for sale:** The change in total salvage and green timber offered for sale each year from all Sierra Nevada national forests, expressed as a percentage of the average amount of timber offered during the six-year period 1994–99 (372 mmbf/year).	Change in timber	−80% −50% 50%
	Forest holes—percentage of forested land (1st decade): Total acres of forest holes created over the next ten years, as a percentage of total forested land in all Sierra Nevada national forests. [A forest hole in this context is a small opening (0.25–2 acres each) created in the forest to facilitate forest regeneration.]	Forest holes	0% 3% 6%

(continued on page 172)

- *outcomes*—projected short-term and long-term changes in habitat and the expected number of acres likely to be burned by wildfires.

For each dimension there are either two or three alternative attributes. Table 9.1 provides a list of the ten attributes within these four dimensions. In our preference elicitation exercise, we measured each attribute at three

TABLE 9.1. *(continued)*

	Attribute description	Attribute name	Levels
Locations	**Percentage of defense zone treated (1st decade):** The total number of defense zone acres treated during the next ten years, as a percentage of total defense zone acres in all Sierra Nevada national forests. [The defense zone is 0.25 mile around state-identified communities and structures. Currently there are 341,352 acres in defense zones.]	Defense zone	75% 85% 90%
	Percentage of threat zone treated (1st decade): The total number of threat zone acres treated during the next ten years, as a percentage of total threat zone acres in all Sierra Nevada national forests. [The threat zone is 1.25 miles beyond the defense zone. Currently there are 2,140,864 acres in threat zones.]	Threat zone	5% 20% 30%
	Percentage of land outside WUI treated (1st decade): The total number of acres treated outside the wildland-urban intermix (WUI) areas (defense and threat zones combined) during the next ten years, as a percentage of total non-WUI acres in all Sierra Nevada national forests. (Currently there are 9,018,897 acres in the national forests, but outside the WUI.)	Land outside WUI	2% 3% 4%

different levels. Hence, these ten attributes, with three measurement levels each, yield 59,049 (3^{10}) unique combinations of potential choices of the levels of treatments, strategies, locations, and outcomes. Not all of these combinations constitute feasible options. For instance, there may be budget shortfalls such that the available funds are not sufficient to mechanically treat 10 percent of the forests and treat another 12 percent through prescribed fires. Some combinations, such as a 50 percent increase in timber offered for sale, may not be compatible with restrictions on the locations where these

TABLE 9.1. *(continued)*

	Attribute description	Attribute name	Levels
Outcomes	**Short-term in change in habitat (1st decade):** Change in old forest habitat acres through fuels treatment or wildfire during the short term (next ten years), as a percentage of total current old forest habitat acres in all Sierra Nevada national forests.	S-T habitat	−1% −2% −3%
	Long-term change in habitat (forty years): Change in old forest habitat acres through growth, fuels treatment, or wildfire in the long term (forty years), as a percentage of total current old forest habitat acres in all Sierra Nevada national forests.	L-T habitat	40% 60% 80%
	Percent change in wildfire acres (forty years): Expected acres burned by wildfires in the fourth decade as a percentage of expected acres burned in the next decade throughout the eleven Sierra Nevada national forests. In the last ten years, wildfires have burned about 70,000 acres per year. It will take at least two decades of fuels treatments before significant changes in wildfire patterns are expected.	Wildfire	5% −20% −50%

treatments might be implemented. Nevertheless, even though many of the combinations either are not technically feasible or do not constitute administratively or politically palatable choices, the set of potentially useful combinations from which to select an overall fire management policy remains large.

Having to sort through and assess all of the almost sixty thousand possible combinations of potential treatments, strategies, locations, and outcomes is not practical. However, it is possible to obtain preferences from an evaluation of the partial set of choices likely to be feasible. Based on a review of the existing management proposals, and in consultation with informed individuals, we developed a set of twenty-three different combinations of attributes and levels that would yield feasible options. The card in figure 9.2

Treatments	
Mechanical	5%
Prescribed fire	8%
Strategies	
Change in timber	-80%
Forest holes	0%
Locations	
Defense zone	90%
Threat zone	20%
Land outside WUI	2%
Outcomes	
Short-term habitat	-1%
Long-term habitat	40%
Wildfire acres	-50%
SCORE	
1-100	
Acceptable	Y N

FIGURE 9.2. Sample card from the card-sort exercise

illustrates an option consisting of a feasible combination of the different levels of treatments, strategies, locations, and outcomes, as described in table 9.1.

The scenario described in the card presented in figure 9.2 combines two treatments—mechanical and prescribed fire—to be applied on 5 percent and 8 percent, respectively, of the total forested acres. The strategies to be employed in this scenario are to have no forest holes and to reduce, by 80 percent, the amount of timber and salvage offered for sale compared to the average amount offered for sale annually over the six-year period from 1994 to 1999. Over the course of a decade, these treatments would cover 90 percent of the defense zone, 20 percent of the threat zone, and 2 percent of the land outside the wildland-urban intermix areas. The expected outcomes of these treatments and strategies would be a 1 percent reduction in the first decade and a 40 percent increase in the fourth decade in old forest habitat acreage. In this scenario, there would also be a 50 percent reduction in the number of acres lost to wildfires in the fourth decade as compared to those lost in the first decade.

We created a deck of twenty-three such cards, each representing a feasible combination of levels of treatments, strategies, locations, and outcomes. We made enough copies of the deck to give one to each of the participants in the workshops. Data on how our respondents ranked the cards in the deck would be sufficient to estimate their preference structures related to the decision dilemma under consideration.

Preferences and Trade-offs

After some introduction and discussion, we asked participants in our workshops to score each card on a scale from 1 to 100, with a higher number signifying a higher level of preference. We also asked them to sort each card as either acceptable or unacceptable if the combination of treatments, strategies, locations, and outcomes reflected on the card actually occurred.

Scoring and sorting the cards is a challenging task. Many respondents found it demanding enough to be a source of some irritation and resistance. However, it also led to strong engagement and high interest. We followed a two-step process in which we first asked the respondents to complete the card-sorting task individually. We next divided them into small groups and asked them to discuss their sorting and scoring criteria and then re-score and re-sort the cards as a group. Having the respondents score the cards individually likely led to more reliable revelation of actual preferences given trade-offs than would have been revealed in typical adversarial participatory meetings or workshops where group dynamics could overwhelm minority perspectives and where people may tend to stake out relatively extreme negotiating positions. Discussion following the sort-and-score exercise provided additional insights into how the respondents interpreted the attributes and combinations of these attributes as they attempted to work through their preferences for the scenarios represented on each card. Going into the exercise, we had assumed that individuals who worked for the Forest Service or other government agencies would be relatively homogeneous in their opinions toward forest management and would hold similar opinions. However, as the ensuing discussions revealed, our initial assumptions about the homogeneity of these groups were incorrect.

We used two different quantitative methods to analyze the data: conjoint analysis and a technique based on Q-methodology. For interested readers, we offer a brief overview of the technical details of these methods in an appendix at the end of this chapter. Here in the main text, it suffices for our purposes to explain that conjoint analysis produces an average preference structure for a group of respondents. By that we mean that conjoint analysis uses the scores assigned to a given card by the respondents to obtain an average utility (or preference rating) for each level of the attributes of the option represented by the card. Thus, with the output of a conjoint analysis, the analyst would be able to determine how an average respondent would score a given card, or estimate how the average respondent would react when presented with a forest management option described in terms of the attribute levels used to create the card.

The second quantitative method is used to analyze the data in a slightly different way. This tool, based on Q-methodology, attempts to identify clusters of individuals with similar views among the respondents. Hence, the purpose of this technique is to identify whether stakeholder groups among the respondents can be identified based on preference structures, rather than on conventional demographic characteristics that are typically used to categorize respondents. In other words, this method is useful in determining whether the average preference structure as identified by the conjoint analysis is representative of all the respondents, or whether there are important subgroupings that would be overlooked if examining only overall averages.

Data Analysis

The card-sort exercise yielded seventy usable decks of twenty-three cards each. Each respondent gave a rating between 1 and 100 to each card in his or her deck. We wanted to determine the preferences of each of the respondents for the different levels of the attributes of the four dimensions—treatments, strategies, locations, and outcomes. Each card represented a feasible management decision scenario. Through this card-sorting process, we obtained an ordering of the twenty-three possible options, together with a score for each of the acceptable options that indicated the strength of the respondent's preference. The conjoint analysis of these data yielded information on the respondents' preferences in the form of the average utility (or preference rating) for each level of the attributes.

Interpreting Conjoint Analysis

Conjoint analysis entails regressing the score (dependent variable) on the attributes (independent variables). The output of the conjoint analysis yields information on the utility (or *part-worth* in the language of conjoint analysis) of each level of the attributes and the importance of each attribute in determining the score for each card, and hence the utility of the attributes. The underlying logic is based on the assumption that in scoring each card, the respondents implicitly assign a utility to each level of the attributes. The score assigned by the respondents to each card is a complex, subjective composite of these utilities. When presented with a sufficient number of these options, it is possible, through the conjoint analysis procedure, to disentangle the utility of each level of each of the attributes, even though the respondents were not explicitly asked to quantify that utility.

The output of the conjoint analysis provides the preference structure of an average respondent who distils the multiple points of view represented in our sample of stakeholders in the forest management planning process. Conjoint analysis also provides additional information about

TABLE 9.2. Preference structure of participants from conjoint analysis

	Attribute and level	Utility	Importance	Total importance
Treatments	Mechanical 1%	5.53	8%	13%
	Mechanical 5%	0		
	Mechanical 10%	8.50		
	Fire 5%	5.26	5%	
	Fire 8%	3.40		
	Fire 12%	0		
Strategies	Timber +50%	12.93	12%	23%
	Timber −50%	0		
	Timber −80%	4.50		
	Holes 0%	0	11%	
	Holes 3%	11.11		
	Holes 6%	0.03		
Locations	Defense 75%	0	3%	15%
	Defense 85%	3.03		
	Defense 90%	1.92		
	Threat 5%	4.07	4%	
	Threat 20%	4.01		
	Threat 30%	0		
	Outside WUI 2%	7.94	8%	
	Outside WUI 3%	5.54		
	Outside WUI 4%	0		
Outcomes	S-T hab −1%	3.73	3%	49%
	S-T hab −2%	0		
	S-T hab −3%	0.34		
	L-T hab 40%	0	24%	
	L-T hab 60%	9.53		
	L-T hab 80%	25.17		
	Wildfire +5%	0	22%	
	Wildfire −20%	22.56		
	Wildfire −50%	23.13		

Note: Attributes as described in table 9.1.

how the respondents weigh or attach importance to each of the attributes in their decision calculus. In assigning a score to a card, the respondents do not view each of the attributes as being equally important. As illustrated in table 9.2, some attributes have a greater weight associated with them than others.

The numbers in the *importance* column in table 9.2 are indicators of the different weight or importance associated with each of the four dimensions in determining the score for each card. In other words, the importance of an attribute may be interpreted as the influence of that attribute in determining the average respondent's overall preference structure. While indicating the strength of the influence of an attribute, the score does not indicate the direction of the preference. For instance, the importance of mechanical means of removing fuels from the forest might be high for one group of individuals because of the strength of their opposition to it, while the importance might be equally high for another group because of the strength of their support for it. The utility they assign to the different attributes indicates the nature or direction of the preference. The utility of each level of an attribute is the contribution of that level of the attribute to the overall utility of that option described by the card.

To help clarify these concepts, table 9.2 lists the different levels of each of the four decision dimensions—treatments, strategies, locations, and outcomes—and, in the last column, shows the importance assigned by the respondents to each of the four dimensions. Almost half, 49 percent, of importance was assigned to outcomes. This 49 percent reflects the importance associated with the three attributes, namely projected short-term effects on habitat (3 percent), projected long-term effects on habitat (24 percent), and projected effects on acreage burned by wildfires (22 percent). Focusing on the long-term habitat attribute, note that there is one level, a 40 percent increase (L-T hab 40%), which is valued the least and therefore assigned a utility of zero. The highest utility value of 25.17 is assigned to the 80 percent increase in habitat forty years hence (L-T hab 80%). Thus, among potential outcomes for this attribute, the respondents, on average, assigned the highest utility to an 80 percent increase and the least to a 40 percent increase in long-term habitat.

Conjoint analysis results can be interpreted as individual utilities for each attribute. Because all the attribute variables in the regression are binary variables, utility is measured on the same scale across all the attributes and can be simply added across the attributes to obtain the overall utility for a card for the average respondent.

Conjoint Analysis Results

Of the attributes that make up the alternative scenarios, the conjoint analysis indicates that the respondents consider the predicted consequences for the forests to be the most important criterion in determining their preferences. From table 9.2 we see that 49 percent of the importance derives from forest outcomes. Further, almost a quarter (24 percent) of the overall importance is associated with long-term outcomes. The strategy employed in managing the forests is important, but to a lesser degree (23 percent). The importance attached to the choice of fuel treatment (mechanical or prescribed fires) is similar to the importance given by the average respondent to the location of the treatments. Among locations, the area outside the wildland-urban intermix is assigned the greatest relative importance (8 percent).

It is difficult to predict the precise consequences of fire management efforts in the forests because of uncertainty about the location and intensity of future fires and the lack of clear evidence linking specific management practices with specific wildfire outcomes. Yet, wildfires and their consequences are a central concern of stakeholders when considering alternate forest management interventions. Perhaps the importance our respondents placed on the effect of forest management plans on fires forty years hence reflects this concern. The long-term effect on habitat (24 percent) and the potential for wildfires forty years hence (22 percent) account for almost half the importance of all the attributes in determining the preferences of the respondents.

Our results indicate that for many respondents the idea of creating forest holes or gaps produces conflicting responses. The creation of gaps can be seen as a positive outcome in that it promotes forest regeneration, but where gaps are created through the use of mechanical treatment some respondents associate them with the collateral damage to the forest this type of treatment may cause. We also noted considerable controversy regarding the harvesting of old-growth stands. Thus, the limit on the diameter of the trees that can be harvested is an important issue for many stakeholders. Larger diameters imply older trees, which environmentalists would want to preserve. Yet loggers, too, assign greater value to trees with larger diameters, because they yield broader and longer planks of wood. We did not explicitly ask the respondents to assign utilities to the diameters of the trees that they would consider appropriate for logging. This point emerged from the discussions. Our initial assumption was that the respondents would be for or against logging. However, the discussions revealed a nuanced attitude toward logging. People seemed to acknowledge that some logging was essential and the debate therefore focused on the size of the trees to be logged rather than the binary choice between logging and not logging. For forest management, we would recommend including this variable in subsequent iterations of the learning network process described in the previous chapter.

Consequently, the strategy employed in managing the forests, represented on the cards by the percentage of the forest opened up for regeneration and the change in the amount of timber and salvage offered for sale, becomes an important consideration. The two strategy attributes, holes and timber, seem to have similar importance (11 and 12 percent, respectively) in the preferences revealed by the respondents.

The remaining perceived importance determining the respondents' scoring of the cards is shared almost equally by the attributes associated with the type of treatment (13 percent) and where it is employed (15 percent). The cards specified two types of treatment: the forest clearing could be achieved either through prescribed fires or by mechanical means. The use of either treatment appears to have a disutility associated with it; however, respondents assigned mechanical treatments relatively higher importance. Note that, as mentioned above, greater importance does not imply a preference for one type of treatment over another. However, the higher importance associated with mechanical treatment (8 percent) means that in deciding what score to assign to a card, the respondents seemed to consider mechanical treatment a more salient factor (positive or negative) than prescribed fires (5 percent).

The three attributes pertaining to where the treatments would occur account for approximately 15 percent of the total importance. One would expect the highest importance to be assigned to the attribute related to the treatments occurring in spaces closer to the built-up areas, that is, in the defense and threat zones, which make up the wildland-urban intermix. However, according to the conjoint analysis, the area outside the wildland-urban intermix seems to be the most important of the three. Note, once again that among the locations for the treatments, the respective importance percentages are 3 percent for the defense zone, 4 percent for the threat zone, and 8 percent for the areas outside the two zones. This interpretation is consistent with the comments heard in several meetings involving participants in the Sierra Nevada Forest Plan Amendment process. Most people seemed willing to acknowledge that protecting lives and structures in the defense and threat zones would require sacrificing other environmental priorities. Thus, the focus of discussion became the level of treatment outside these areas; hence the highest importance (8 percent) being assigned to areas outside the wildland-urban intermix.

Simulation Using Conjoint Results

Although we asked respondents to score only twenty-three different combinations, conjoint analysis provides the average utility for each level of the ten attributes, making it possible to estimate an average utility score for each of the 59,049 possible combinations of the three levels of the ten attributes. We note here that conjoint analysis, like any statistical procedure, has its limitations. We discuss these briefly in the technical appendix.

As mentioned earlier, the score assigned to each card is a subjective composite of the utilities associated with each level of the attributes. An attractive feature of conjoint analysis is that it *unpacks* the score to produce estimates of the utilities assigned to each level of the attribute from the scores assigned to a limited number of cards. Knowledge of the utility of each attribute level allows us to estimate scores for new options that the respondents did not consider. The key point is that in an iterative, deliberative, learning network process, it would be possible using this approach to return to stakeholders with new information that had not previously been considered about participants' attitudes toward the various options. These insights might serve as the basis for some movement toward agreement that otherwise would be overlooked. This ability resulting from conjoint analysis to combine attributes to create new options has been used with demonstrated success in the marketing context to develop desirable new products and services (Shocker and Srinivasan 1977; Zinkhan, Holmes, and Mercer 1997). Our proposed process for managing wicked problems supplements the learning network approach described in the literature (National Research Council 1996) by providing participants with new information regarding preferences for policy alternatives constructed from other combinations of attributes that may be valuable in advancing the discussions and helping the stakeholders assess new, heretofore unexplored options.

It is possible to construct an estimated score for every possible combination of attribute values by using the parameter estimates from the conjoint analysis. As noted previously, there are 59,049 such possibilities. Recall that our source data for these simulations are the respondent scores assigned to the twenty-three cards, which yielded the utilities and importance values shown in table 9.2. These scores ranged from 1 to 100. We calculated the scores for each of the 59,049 potential options using the coefficient estimates from our sample. Options with the highest estimated scores are seen as the most desirable by the average stakeholder. Thus, it is possible to take, say, the scores of the top twenty options and look for com-

TABLE 9.3. Top twenty options identified from simulation

	Attribute	Option description
Treatments	Mechanical—percentage of forested land (1st decade)	2 options call for 1% 18 options call for 10%
	Prescribed fire—percentage of forested land (1st decade)	15 options call for 5% 5 options call for 8%
Strategies	Change in timber and salvage offered for sale	All options call for 50% increase
	Forest holes—percentage of forested land (1st decade)	All options call for 3%
Locations	Percentage of defense zone treated (1st decade)	1 option calls for 75% 14 options call for 85% 5 options call for 90%
	Percentage of threat zone treated (1st decade)	11 options call for 5% 9 options call for 20%
	Percentage of land outside WUI treated (1st decade)	6 options call for 2% 4 options call for 3%
Outcomes	Short-term change in habitat (1st decade)	All options call for –1%
	Long-term change in habitat (forty years)	All options call for 80%
	Percent change in wildfire acres (forty years).	8 options call for –20% 12 options call for –50%

Note: Attributes as described in table 9.1.

monalities among these options. This is what we mean by a simulation in this context. The technique allows us to simulate respondents' preferences for a much larger range of alternative policies and outcomes than would be possible in workshops, town meetings, or other similar fora.

Examining the top twenty options thus identified reveals the features shown in table 9.3. Each of these options consists of some combination of the levels of the attributes. Hence, an *All* in the last column of table 9.3 implies that the top twenty options all include the attribute at that level.

As tempting as it may be to see these results as providing immediately useful guidance for policy makers, there are two issues that deserve attention. First, the simulation process does not account for whether the highly rated options are feasible in the sense of representing implementable options. This assessment would have to be done separately. Second, and more important for our discussion, conjoint analysis produces an average. This single aggregate preference ordering may have limited value in a complex decision environment. In a marketing setting, it may be useful to assess the preferences of a typical consumer, but in a wicked problem setting, focusing on the average stakeholder masks both the extent of disagreement among stakeholder groups and the presence of potentially powerful minorities who may strive to undermine implementation of any strategy they find unacceptable, even if it is acceptable to the average stakeholder. For example, in table 9.3, eighteen of the top twenty options entail increasing mechanical thinning to 10 percent of all forestlands. Expanding timber harvesting by 50 percent is also a feature of all twenty options. If implemented, however, this would likely result in numerous lawsuits from those opposed to timber harvesting even if the option had both substantial public support and the potential to increase threatened natural habitat over the long term.

Thus, those engaged in the learning network that we propose need to understand not just an aggregation of public preferences but also the preferences of key stakeholder groups. One option for doing so might be to identify the groups in advance based on institutional affiliation, employment, or some other objective criteria, as we did in conducting our questionnaire. But such categorizations are likely to be flawed. For example, some Forest Service employees may favor increasing timber production, while others may prefer greater conservation. Grouping all Forest Service employees together would thus mask important differences in value positions (Martin and Steelman 2004). Further analysis of the data from the card-sort exercise is necessary to reveal these within-group differences.

To summarize, conjoint analysis yields information regarding average preferences for the respondents as a group. By definition, these averages mask the full range of variation in the preferences of individual respondents. In order to explore this range of preferences more fully, an alternative approach would be to group individuals based on subjective assessments provided by respondents themselves. A method that offers precisely such a grouping is Q-type analysis.

Q-type Analysis

Unlike the analysis underlying conjoint analysis, which assumes that the score for each card depends upon the attributes represented on each card, Q-type analysis groups the individuals into categories depending upon how they scored the cards. Whereas conjoint analysis provides information on how an average respondent would choose among the various cards, we also want to know whether there are clusters of individuals who have similar preferences and, if so, how many clusters there are and what the preference structures are for these subgroupings. The unit of analysis therefore, is no longer the cards from which we want to estimate the underlying average preference structure. Instead, the unit of analysis is the individual respondent, and the objective is to determine whether respondents can be clustered into groups of like-minded individuals with similar preferences, as indicated by their scoring of the cards. Hence, for Q-type analysis, we focus on finding clusters among the individuals or the rows of the data matrix, rather than looking for the relationship between the scores and the attributes, which make up the columns of the data matrix. So, for the Q-type analysis we work with the same data but start with a modified version of the data matrix that we used for conjoint analysis. The details of the modification are provided in the appendix.

Q-type Analysis Results

The statistical procedure underlying the Q-type analysis is a factor analysis, and its output is to be interpreted as the grouping of our seventy respondents according to their subjective assessments of the twenty-three cards. As with many statistical procedures, factor analysis entails an analysis of the variation in the data. The technique seeks factors or underlying dimensions in the data that capture the majority of the variation observed in the data. The technique seeks factors that maximize the amount of variation accounted for, while trying to find factors distinct from each other so as to minimize the overlap among them. Hence, in our analysis the factors represent aggregates or groups of respondents who are similar in their assessments of the treatments, strategies, locations, and outcomes of the forest management plans but differ in those assessments from people in other groups. In other words, we are grouping the respondents by their value orientations expressed in their evaluation of the options represented in the twenty-three cards. In this way, we are no longer grouping the respondents by their employment or group membership. In what follows, we shall use the term *factors* to indicate the output of the computations that sort the respondents into groups of individuals with similar preferences as indicated by the scores they assigned to the cards.

The computational analysis yields twenty-three unique factors. The underlying mathematical procedure attempts to reduce the number of factors to the smallest number that accounts for the largest amount of variation in the data. Table 9.4 shows factors A, B, and C, which numerically describe the value orientations of three like-minded groups of respondents. The bottom row of table 9.4 shows that these three factors capture 87.2 percent of the total variation, with factor A accounting for 64.5 percent, factor B accounting for 18.1 percent, and factor C accounting for 4.6 percent. Thus, we need only three groupings of the respondents corresponding to these three factors, to capture almost 90 percent of the variation in the preferences observed among the respondents.

Thus, each factor represents a group of respondents who share a similar preference structure. It is useful, as shorthand, to consider these factors as value orientations. To see how the groups vary in their outlooks, it is possible, from the output shown in table 9.4, to estimate the scores that an individual in a group would assign to a card. Differences in the resulting average scores reflect differences in each group's views, on average, of each attribute. For example, table 9.4 reports on these averages for the forest management example used here.

To clarify how the numbers in the table should be interpreted, for each cell a value of 0 represents the mean for that attribute and level, and the actual number reported in a particular cell represents the distance from the mean of the results for that cell in standard deviation units. These entries are referred to as *factor loadings*. For example, consider −0.059, which is the entry for "Fire 5%" in the treatments row and factor A column. This number indicates that among individuals with value orientation A, the average score assigned to all options that included prescribed burning on 5 percent of forestlands was slightly below the average of 0. Now consider the entry (factor loading) on the same row for factor C, which is 0.276. This number is interpreted as being about one-quarter of a standard deviation above the mean. Thus, respondents with value orientation C would appear to have, relatively speaking, a more positive attitude toward the treatment option of Fire 5% than respondents with value orientation A.

We use shading in table 9.4 to indicate those cells for which the entry (factor loading) is at least ±0.2 to highlight the attributes that play a dominant role in defining the value orientation. In some instances the attribute is

TABLE 9.4. Q-type analysis means

	Attribute	Factor A	Factor B	Factor C
Treatments	Mechanical 1%	−0.402	0.310	−0.207
	Mechanical 5%	0.091	0.022	0.298
	Mechanical 10%	0.312	−0.333	−0.091
	Fire 5%	−0.059	−0.181	0.276
	Fire 8%	0.170	0.356	−0.154
	Fire 12%	−0.112	−0.175	−0.122
Strategies	Timber +50%	0.587	−0.474	−0.024
	Timber −50%	−0.164	0.301	0.194
	Timber −80%	−0.423	0.173	−0.170
	Holes 0%	−0.156	0.788	−0.187
	Holes 3%	0.009	−0.303	0.327
	Holes 6%	0.147	−0.485	−0.140
Locations	Defense 75%	0.042	−0.019	0.118
	Defense 85%	0.011	0.011	−0.007
	Defense 90%	−0.053	0.007	−0.112
	Threat 5%	−0.259	−0.074	0.052
	Threat 20%	0.013	0.160	−0.052
	Threat 30%	0.246	−0.086	0.000
	Outside WUI 2%	−0.168	0.169	−0.142
	Outside WUI 3%	−0.029	0.017	0.161
	Outside WUI 4%	0.198	−0.186	−0.019

shaded for all three factors. In interpreting the sign, a positive value implies that the respondents have a positive preference for that attribute, whereas a negative value implies that the attribute plays a negative role in determining the preferences of the respondents in that group. For instance, regarding the use of mechanical treatments, the respondents with value orientation

TABLE 9.4. *(continued)*

	Attribute	Factor A	Factor B	Factor C
Outcomes	S-T hab –1%	–0.143	–0.055	0.187
	S-T hab –2%	0.124	0.076	–0.250
	S-T hab –3%	0.019	–0.021	0.064
	L-T hab 40%	–0.386	0.181	–0.214
	L-T hab 60%	0.196	0.123	0.166
	L-T hab 80%	0.190	–0.304	0.049
	Wildfire +5%	–0.247	–0.244	0.596
	Wildfire –20%	0.034	0.119	–0.124
	Wildfire –50%	0.214	0.125	–0.472
Variation accounted for		64.5%	18.1%	4.6%

Note: Attributes as described in table 9.1.

described by factor A are in favor (0.312) of a 10 percent level of mechanical treatment and are opposed (–0.402) to a 1 percent level of mechanical treatment. The group identified by factor B, on the other hand, favors (0.310) a 1 percent level of mechanical treatment and is opposed (–0.333) to a 10 percent level of mechanical treatment. The third group, described by factor C, favors (0.298) a 5 percent level of mechanical thinning, seems to be opposed (–0.207) to a 1 percent level of mechanical treatment, and is perhaps indifferent (–0.091) to a 10 percent level of mechanical treatment.

Respondents with value orientation A, which accounts for approximately 65 percent of the variation observed among the respondents, appear most concerned with nine attribute values (the cells shaded in the Factor A column). People in this group are most favorable toward increasing timber production (Timber +50%: 0.587) and more expansive use of mechanical thinning (Mechanical 10%: 0.312). They are also favorably inclined toward more extensive treatment in the threat zones (Threat 30%: 0.246) and want substantial reductions in acreage lost to wildfires (Wildfire –50%: 0.214). On the other hand, the shaded negative numbers in the Factor A column suggest that respondents with that value orientation are opposed to substantial reductions in timber harvests (Timber –80%: –0.423), minimal use of mechanical thinning (Mechanical 1%: –0.402), and modest gains in habitat over the long term (L-T hab 40%: –386). These respondents would prefer more substantial (60% and 80%) gains in long-term habitat as indicated by the positive scores associated with those attribute levels.

Individuals with value orientation B appear to be more concerned with treatments and strategies than with locations or outcomes. While people in this group are opposed to increased incidence of wildfires (Wildfire +5%: –0.244), and paradoxically, to large increases in long-term habitat gains (L-T hab 80%: –0.304), they strongly prefer that no forest gaps be created (Holes 0%: 0.788), that timber harvests not be expanded (Timber +50%: –0.474 and Timber –50%: 0.301), that mechanical treatment be minimized (Mechanical 1%: 0.310 and Mechanical 10%: –0.333), and that prescribed burns be used in a middle range (Fire 8%: 0.356) of the alternatives for forest areas to be treated.

Individuals with value orientation C seem more inclined to prefer moderate mechanical thinning (Mechanical 5%: 0.298) and forest gaps (Holes 3%: 0.327) and limited use of prescribed fires (Fire 5%: 0.276). Surprisingly, they appear willing to accept increased wildfire losses (Wildfire +5%: 0.598).

There appears to be little concern among respondents in all the three groups about where the treatments occur. Issues of location may be outweighed by concerns about treatments and strategies. Although outcomes appear to be of some interest, it seems that the respondents acknowledge the uncertainties involved in projecting outcomes in the short term as well as the long term, and these data seem to suggest, counterintuitively, that projected outcomes matter less than the processes employed in achieving them. This apparent preference for process over outcomes conforms to our earlier discussion that the primary risks for the Forest Service may be associated with political processes rather than projected ecological outcomes.

Simulation Using Q Results

The Q-type analysis results can also be used to simulate responses to all 59,049 possible combinations of the three levels of the ten attributes, and to thus estimate the expected evaluations of these combinations from each group. This computation would be tantamount to having the respondents in the groups score each of the 59,049 cards. Because the factor scores are standardized, the ratings can be interpreted as the expected standardized evaluation by each group for each of these 59,049 options.

As with the conjoint analysis simulation, this process would not account for scientific or political feasibility.

From a statistical perspective, it would be appropriate to wonder how much weight to place on the preferences of the respondents with value orientation C. Of the approximately 90 percent of the variation in the preferences jointly accounted for by these three factors, factor C accounts for as little as 5 percent. Therefore, for ease of exposition and because of the small amount of variation in the data that is accounted for by factor C, we focus below only on factors A and B, which jointly account for about 83 percent of the total variation in preferences.

TABLE 9.5. Seven potential compromise options

	Attribute	*Option description*
Treatments	Mechanical—percentage of forested land (1st decade)	All call for 8%
	Prescribed fire—percentage of forested land (1st decade)	All call for 5%
Strategies	Change in timber and salvage offered for sale	All call for 50% increase
	Forest holes—percentage of forested land (1st decade)	All call for 0%
Locations	Percentage of defense zone treated (1st decade)	3 call for 75% 4 call for 85%
	Percentage of threat zone treated (1st decade)	5 call for 20% 2 call for 30%
	Percentage of land outside WUI treated (1st decade)	4 call for 2% increase 2 call for 3% increase 1 calls for 4% increase
Outcomes	Short-term change in habitat (1st decade)	All accept 2% loss
	Long-term change in habitat (40 years)	All call for 60% gain
	Percentage change in wildfire acres (40 years)	All call for 50% reduction in acres burned

Note: Attributes as described in figure 9.1.

In an applied policy context, public managers would seek options that have some potential to be acceptable to individuals in these two key groups. In examining the almost sixty thousand options, we find that seven are near the highest levels of acceptance for both value orientations and no other options are preferred over these seven. These are the options most likely to be acceptable to respondents with either value orientation A or B. The attribute levels that constitute these seven combinations are listed in table 9.5.

Here again it is tempting to rely on findings such as those in table 9.5 to design options likely to appeal to respondents with either value orientation A or value orientation B. There appears to be agreement on desired outcomes, treatments, and strategies. Differences remain only around locations, so that would seem to be where future negotiations should focus. If this were true, it would suggest a good deal of optimism is in order about the possibility of resolving long-term differences. But here again, the technique masks important information.

Recall that among the seventy participants in our card-sort activity the Q-type analysis suggests that there are three largely nonoverlapping value orientations defined by the factors A, B, and C, shown in table 9.4. Consider the information shown in figure 9.3. In this figure, given the limited

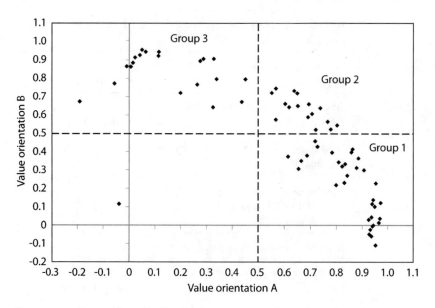

FIGURE 9.3. Respondent distribution between two value orientations

amount of variation in the data that is accounted for by factor C, and the restrictions imposed by the two dimensions of a flat page, we ignore value orientation C and plot each of the seventy respondents on the basis of their scores along value orientations A and B. We divide the graph into quadrants by inserting vertical and horizontal lines at 0.5.

As is apparent in the graph, while there is a cluster of individuals to the left of 0.5 on the horizontal axis and above 0.5 on the vertical axis (group 3) and another to the right of 0.5 on the horizontal axis and below 0.5 on the vertical axis (group 1), there are also quite a few people in the northeast quadrant (group 2) who are in neither camp. Thus, whether due to ambivalence or an honest sharing of values from the value orientations A and B, the figure suggests that the two value orientations actually represent the poles of a continuum. In plotting the respondent scores, we obtain an arc of data such that the points seem to scatter from near the horizontal axis beginning with people in group 1, and progress through groups 2 and 3, going slightly past the vertical axis. Individuals in group 1 are those who score higher on value orientation A and individuals in group 3 score higher on value orientation B. As the figure suggests, however, dichotomizing the participants' value positions masks the complexity of the underlying values.

A "Wicked" Analytic Proposal

What we see then is that in the context of wicked problems conjoint analysis has the strength of offering powerful models to capture aggregate preferences, but runs the risk of masking potentially important differences among powerful minorities. Q-type analysis is useful in identifying value orientations, but runs the risk of imposing artificial distinctions that may in fact distort the actual distribution of responses to new potential options.

Because both approaches offer potentially important insights yet suffer from shortcomings in the present context, we propose a hybrid approach that makes use of both techniques. In this approach, we first carry out Q-type analysis to identify distinct value orientations and potential groupings. Using this information we place each respondent in a relatively homogeneous group based on self-declared preferences as revealed in the card sort. Thus, we have the individuals in groups 1, 2, and 3. We return to the original card sort data and use the three groupings shown in figure 9.3 to conduct three separate conjoint analyses of the scores given by each respondent to the deck of twenty-three cards.

Results

The results of this hybrid effort are shown in table 9.6. As can be seen in the table, there are substantial differences between the groups in their assessments of various option attributes. The story that emerges from table 9.6 is quite different from the one told in table 9.2. In retrospect, from looking at figure 9.3, it stands to reason that table 9.2 would most closely resemble group 2 because the overall conjoint analysis is the average of the seventy respondents who are distributed in and around group 2. Respondents in group 2 place over 50 percent of the importance on outcomes, which accounts for about 45 percent of the importance in the aggregate (as illustrated in table 9.2). Strategy is second highest in importance for all the groups.

However, table 9.6 indicates that the options that would be attractive to group 2 would not be attractive to people in the other groups. Consequently, relying on the overall conjoint analysis would yield solutions that might seem attractive to the *average* respondent, but would not yield implementable solutions. Respondents in groups 1 and 3 place substantial importance on harvesting timber and salvage (group 1: 36 percent; group 3: 47 percent). While members of both groups indicate that the role of timber harvesting is important in determining their preferences regarding strategies, their preference structures are almost polar opposites in that one group views timber harvesting in a positive light and the other negative.

It would appear that group 1 would welcome increases in logging whereas group 3 would seek substantial reductions in timber harvesting and salvage. Hence, the moderate importance placed in the aggregate on the strategy attributes masks the diametrically opposed views held by respondents in groups 1 and 3, both of whom consider sales of timber to be of great import, but for different, opposing, reasons.

A look at the preferences would seem to suggest that all three groups prefer that 3 percent of the forest be opened to assure regeneration. However, while the preference structures in groups 1 and 2 seem to be similar in that members of these groups see some utility in having more open spaces, those in group 3 appear to favor having no forest holes.

Hybrid Simulation

In chapter 6, we proposed that preference approval voting as described by Brams and Sanver (2009) might provide an approach to converting preference information into scores that would allow us to find a preferred option. In conducting the card-sort exercise (asking the participants to first identify those combinations that they considered acceptable, and to then assign a score between 1 and 100 to those options that were deemed acceptable), we collected the information necessary to implement the Brams and Sanver (2009) preference approval voting procedure.

With the information summarized in table 9.6, it is possible to return to the 59,049 possible combinations of attributes and attempt to predict how each of the groups would evaluate each option. We compute the scores that members of each of the three groups would give these combinations. Table 9.7 reports the descriptive statistics for the predicted rating scores for each group.

A naive statistical analysis of the scores can be misleading. Not surprisingly, the predicted scores for group 1 and group 3 are negatively correlated, but not overwhelmingly so ($r = -0.305$). The expected scores from group 2 are positively correlated with the scores from both of the other two groups. With group 1, the correlation is 0.714 and with group 3, 0.347. However, as illustrated by the results presented in figure 9.3 and table 9.6, it would be misleading to assume that groups 1 and 2 have similar preference structures and that a compromise solution can be attained without much effort.

Once again, the search is for common ground on which to build potential agreement. In this instance, we selected all options with predicted scores above the ninetieth percentile for each group. Of the 59,049 potential combinations, only twenty-eight options satisfied this criterion. Of the twenty-eight, six were not dominated by some other option in the set. These six options are reported in table 9.8, along with the predicted ranking from each group. In the table, multiple attribute values are reported only when the options differ in a given attribute value.

It is apparent from the analysis thus far that these six options will be somewhat controversial. For example, all six require further reductions in timber harvesting. Clearly group 1 will not like this option. On the other hand, all options require maximal use of mechanical thinning, which will not please group 3. There is still a good deal of negotiating to be done among the groups around these options. The point is that each group should find something attractive in each option because these options are selected based on congruence with the exhibited value orientations.

Feasibility

The next steps would be first to ascertain whether these six options are feasible; second to return to the participants in the established learning network to verify that the value orientations and revealed preferences are valid; and third to engage in the next round of discussions based on these six options.

TABLE 9.6. Q-type conjoint analysis: Results from the hybrid approach

	Attribute and level	Group 1 (N=33)		Group 2 (N=16)	
		Utility	Importance	Utility	Importance
Treatments	Mechanical 1%	7.10	9.9%	0	7.7%
	Mechanical 5%	0		5.9	
	Mechanical 10%	14.7		9.9	
	Fire 5%	8.59	5.8%	3.2	2.8%
	Fire 8%	6.60		3.7	
	Fire 12%	0		0	
	Total importance		15.7%		10.5%
Strategies	Timber +50%	39.8	26.9%	8.0	6.3%
	Timber −50%	0.1		0	
	Timber −80%	0		6.8	
	Holes 0%	0	9.1%	0	9.1%
	Holes 3%	13.5		11.7	
	Holes 6%	0.4		2.2	
	Total importance		36.0%		15.4%
Locations	Defense 75%	0	3.2%	0	3.7%
	Defense 85%	4.8		4.8	
	Defense 90%	2.5		0.7	
	Threat 5%	4.4	4.3%	4.8	4.1%
	Threat 20%	6.3		5.2	
	Threat 30%	0		0	
	Outside WUI 2%	4.4	3.2%	9.0	7.0%
	Outside WUI 3%	4.7		5.2	
	Outside WUI 4%	0		0	
	Total importance		10.7%		14.8%
Outcomes	S-T hab −1%	7.35	4.9%	4.0	3.1%
	S-T hab −2%	0		0	
	S-T hab −3%	2.51		1.4	
	L-T hab 40%	0	20.4%	0	27.9%
	L-T hab 60%	8.4		14.0	
	L-T hab 80%	30.2		35.8	
	Wild fire +5%	0	12.3%	0	28.3%
	Wild fire −20%	18.0		31.1	
	Wild fire −50%	18.2		36.3	
	Total importance		37.6%		59.3%

Note: Attributes as described in figure 9.1. The observation in the bottom left-hand corner of figure 9.3 is excluded from the analysis (thus, N=69).

Group 3 (N=20)	
Utility	Importance
0	5.9%
0	
0	
0	2.4%
2.1	
1.0	
	8.3%
0	32.1%
27.3	
27.8	
8.0	14.9%
12.9	
0	
	47.0%
3.4	4.3%
0	
3.7	
4.0	4.6%
0.3	
0	
9.4	10.9%
8.0	
0	
	19.8%
2.6	3.0%
2.3	
0	
0	2.8%
1.5	
2.4	
0	19.1%
16.6	
16.5	
	24.9%

These discussions would not be easy. Even though these six options have many attribute levels in common, their rank ordering, shown at the bottom of table 9.8, suggests that the preferences for these attributes are at odds with each other across the groups. (In the language of Game Theory, however, there does appear to be a Condorcet winner in this set, if the groups are assumed to be equal in size and all choices are made by pair-wise voting. Under such conditions, option 1 would emerge as preferred over the other five options.)

Limitations

From a purely technical perspective, we have taken a data matrix and attempted to extract information from it in two different ways. We learn that both approaches have the potential to provide valuable information. In fact, they provide complementary information.

Conjoint analysis is a potent tool for gathering data on choices and trade-offs among attributes and for analyzing these data to extract preference structures. Unfortunately, it provides an average preference structure. Having a sample representative of all the opinions in the population is not sufficient for obtaining policy-relevant information because in this instance aggregate information is not adequate, and quite likely misleading.

The usual strength of conjoint analysis, which has been particularly useful in marketing contexts, is its ability to extract the set of attributes that would be most attractive to the largest group of consumers. In our context of a management plan for the Sierra Nevada national forests, however, minority opinions and preferences have the potential to derail any solution crafted to appeal to the *average* stakeholder.

TABLE 9.7. Descriptive statistics on utilities by group

Group	N	Mean	St. Dev	Min	Max
1	33	67.50	26.44	0	148.1
2	16	67.77	23.60	0	128.3
3	21	52.35	17.00	0	86.7
All	70	54.08	17.78	0	104.9

Provided the sample of stakeholders is representative of the full range of opinions prevalent on the subject, Q-type analysis has the potential to identify the various value orientations. Conjoint analysis of the data along each of these value orientations can yield information that could lead to compromises and potentially implementable solutions.

Engaging in public participatory processes entails a long-term commitment of time and often becomes an onerous and frustrating activity for participants. Hence, by definition, the persons who generally participate in such processes for extended periods are self-selected and likely constitute a biased sample of stakeholders. Consequently, it is not always feasible to obtain the full spectrum of opinions that might need to be considered when selecting forest management plans.

Another limitation is the design of the data collection instrument itself. Conjoint analysis is usually used in contexts where respondents are asked to rank or score a variety of realistic alternatives. Providing such information is demanding, making the data collection complex, time consuming, and expensive.

Results are affected by the choices offered. In the case of wicked problems, the number of attributes required to adequately describe alternatives can quickly become quite large. Even with well-formulated experimental designs, the choices necessary for the stakeholders to provide sufficient information to develop a full preference profile becomes prohibitively large.

Conclusions

Regional foresters are required by law to involve the public in decisions regarding the management of the national forests. This decision context seems ideal for implementing the deliberative analytic process proposed by the National Research Council (1996). We set out to determine how the

TABLE 9.8. Attributes and rank ordering of six options

	Attribute and Level	Option					
		1	2	3	4	5	6
Treatments	Mechanical 10%	Y	Y	Y	Y	Y	Y
	Fire 5%	Y			Y	Y	Y
	Fire 8%		Y	Y			
Strategies	Timber –80%	Y		Y	Y		Y
	Timber –50%		Y			Y	
	Holes 3%	Y	Y	Y	Y	Y	Y
Locations	Defense 85%	Y	Y	Y	Y	Y	
	Defense 90%						Y
	Threat 5%	Y	Y		Y	Y	Y
	Threat 20%			Y			
	Outside WUI 2%	Y	Y	Y			
	Outside WUI 3%				Y	Y	Y
Outcomes	S-T hab –1%	Y	Y	Y	Y	Y	Y
	L-T hab 80%	Y	Y	Y	Y	Y	Y
	Wild fire –50%	Y	Y	Y	Y	Y	Y
Ranks	Group 1	3	5	4	2	1	6
	Group 2	2	4	1	3	6	5
	Group 3	3	2	5	4	6	1

Note: Attributes as described in figure 9.1. Y indicates the level of the attribute is included in the option.

regional forester responsible for the Sierra Nevada national forests in California, if engaged in such a learning network process, could obtain credible information from multiple stakeholders and use that information to develop an implementable forest management plan. We can claim limited progress toward our goal. We have clearly not tamed the wicked problem, but we have developed a process that can yield valuable, previously untapped information on stakeholder preferences. This new information in turn can be used to begin or strengthen an informed conversation about value conflicts and other uncertainties. We have also made progress toward providing analytical support that can help stakeholders understand their values and preferences and the potential consequences of those values and preferences for real-world policy outcomes.

Conjoint analysis and Q-methodology have demonstrated their utility in a variety of contexts. Our question is whether they have a useful role to play in analyzing data and informing decisions in the context of inherently complex and apparently intractable policy dilemmas. Our answer is *no* if only one tool is used, but *perhaps* if they are used together in the innovative ways we have described.

Q-type analysis allows us to tease out different value orientations present among the stakeholders. Conducting conjoint analyses for each of these value sets provides a spectrum of preference structures that could be used to inform participants in learning networks and support the decision-making process.

In our analysis of forest management preferences, we find that long-term consequences are important, but different value orientations place different importance on the attributes that define the consequences. Similarly, we also discover that strategies matter, but that preferences regarding these strategies are contradictory among the two value orientations we studied.

We find that the analytic deliberative participatory process proposed for addressing wicked problems can be supported by the use of conjoint analysis and Q-type analysis. The greatest advantage these two techniques offer, when implemented in concert, is their ability to obtain preference structures based on a limited set of choices. Q-type analysis allows us to identify multiple value orientations and conjoint analysis provides us with the information to generate preferred options. These options are potentially viable solutions that can be presented to stakeholders as the foundation upon which to build implementable management plans.

Technical Appendix

Ever since Lancaster (1966) suggested that when individuals choose, they have preferences for the attributes of a product but not the product itself (Fishbein 1967), researchers have designed approaches to decision making that attempt to assess individual preferences in terms of attributes rather than holistic assessments. Thus, in our case, where the stakeholders are trying to choose among forest management options, the choice according to Lancaster (1996) would be based on how people evaluate each option on the basis of the different dimensions—treatments, strategies, locations, and outcomes—and not on the basis of some overall comparison of the possible managerial actions.

Researchers from various disciplines have developed a number of techniques for eliciting and analyzing data on human perceptions and preferences. These include conjoint analysis (Luce and Tukey 1964; Green and Srinivasan 1990), analytic hierarchy process (Saaty 1980), Q-methodology (Thompson 1935; Stephenson 1935, 1953; Block 1961), Delphi method (Pill 1971), and nominal group technique (Delbeq and Van de Ven 1971). Data collection for these techniques is based on obtaining subjective information regarding the respondents' preferences. The techniques that we used in our study, conjoint analysis and Q-methodology, both have their origins in the psychology literature.

Conjoint analysis involves the measurement of psychological judgments, such as preferences, acceptable thresholds, or perceived similarities or differences among options. Conjoint analysis is a technique in which respondents are given various options for which they express their preferences. The researcher selects the options in advance to incorporate the relevant range of attributes. Each option thus consists of a specific set of attributes and each option's description includes information on these attributes. Thus, the cards that the respondents sorted in our study describe options in terms of the values of the levels of each of the four dimensions—treatments, strategies, locations, and outcomes. The responses provide an implicit snapshot of each respondent's preference structure across the range of attribute values and combinations of the attributes. By analyzing these responses, it is possible to create for each individual an aggregate preference structure that will provide insights into not only the subjective judgments regarding the choices the respondents were offered but also regarding various other combinations and levels of these attributes. Hence, in scoring the cards, the respondents provided information on their preferences for the combinations of the different levels of attributes represented by each card. Analysis of these data through conjoint analysis yields the information we are seeking regarding how these individuals would score any option, not just the twenty-three cards that they actually scored.

Since the early 1970s, conjoint analysis has been used extensively in marketing and business contexts for measuring the trade-offs consumers make in choosing among products, services, or service providers (Green and Srinivasan 1978, 1990). The technique has also been used in the development of new products and services based on the preferences for attributes that are considered desirable (Shocker and Srinivasan 1977). The use of conjoint analysis is not restricted to business and marketing decisions. It has been applied in understanding the preferences of negotiators (Greenhalgh and Neslin 1981), in identifying trade-offs between competing values in public opinion (Shamir and Shamir 1995), and in assessing healthcare alternatives (Ryan and Farrar 2000). More relevant to the current context, other researchers have used the technique to design a forest park (Zinkhan, Holmes, and Mercer 1997) and to study forest products (Reddy, Bush, and Roudik 1995).

Unlike conjoint analysis, which has spawned a small consumer research industry, Q-methodology is a relatively sparingly used data analysis tool (McKeown and Thomas 1988). Although not frequently applied, it has been used in studying a wide range of topics, including organizational culture (O'Reilly, Chatman, and Caldwell 1991), democratic theory (Dryzek and Berejikian 1993), citizenship and public participation (Theiss-Morse 1993), participant perspectives in forest management (Steelman and Maguire 1999), and agency values and objectives in land management (Martin and Steelman 2004). Statistical procedures are generally associated with the positivist approach to research, where one of the important assumptions is that the data are objective measures of the phenomena being studied. Postpositivist approaches do not assume that data are necessarily objective; in fact, in many instances, the purpose of the research is to unearth subjective perspectives. Emphasizing the use of subjective data in Q-methodology, Durning (1999) has proposed its use as a bridge from positivist to postpositivist policy analysis (Durning and Edwards 1992). Brown, Durning, and Selden (1999) go further to provide a comprehensive overview of the use of Q-methodology for public policy analysis.

Durning (1999) offers five specific suggestions for using Q-methodology in the public policy realm. The rationale for applying Q-methodology in our case study of Sierra Nevada national forest management is expressed in his fourth suggestion, where he writes: "Q-methodology can be used as an alternative to survey research to investigate the attitudes of different groups toward a policy proposal. Alternatively, it can be used to help a group work together to select from among different options, providing insights into different attitudes that can advance a group decision-making process" (Durning 1999, 406).

Formal Models

The estimation procedures used for both conjoint analysis and Q-methodology are commonly used statistical methods. Conjoint analysis requires, in the main, the estimation of regression coefficients. Q-methodology employs factor analysis.

Conjoint Analysis

Conjoint analysis provides two types of information. It estimates the respondent's *part-worth* or *utility* for each level of an attribute. The logic underlying the technique is that that when individuals choose among alternatives the choice is based on the attributes that make up the choices. So if we can determine how each attribute is valued—the utility of each attribute—then we can ascertain the overall utility of a choice by simply combining the individual utilities of the attribute levels that make up that choice. In addition, conjoint analysis also yields the importance of each attribute in determining the respondent's overall preference for a card, which describes a potential decision scenario.

So that there is a common understanding of the choices and how they are to be evaluated, the conjoint analysis approach to analyzing preferences for options requires that the attributes, experimental design, data collection, estimation, and interpretation are clarified by the researchers for the respondents at the outset. Respondents are asked to assign preferences, often in the form of a score or a rank, to options. Each option is represented in terms of its attributes occurring at a particular level. The estimation of the utility (part-worth) of each level of the attribute is done typically through a model of the form

$$P_{ij} = \sum_k \sum_m \mu_{ikm} a_{jkm}$$

where

P_{ij} is the score of individual i for option j;
μ_{ikm} is the part-worth or utility of individual i for level m of attribute k;
a_{jkm} is a binary variable indicating the absence or presence of level m of attribute k in option j.
$1 \leq i \leq N$, number of respondents;
$1 \leq j \leq J$, number of options;
$1 \leq k \leq K$, number of attributes;
$1 \leq m \leq M(k)$, number of levels of attribute k.

The statistical procedure used to estimate the part-worths depends upon the data collection procedure. For rank-ordered data, monotone analysis of variance is the best tool; for interval-scaled data, ordinary least squares regression is used to estimate the part-worths.

The importance of an attribute in determining the preferences is obtained by dividing the range of the part-worths for the attribute by the sum of all part-worth ranges. Importance is often expressed as a percentage.

Q-Type Methodology

Q-methodology is more than a statistical data analysis tool. It is an approach that includes well-defined procedures for data collection and data analysis. Focusing on the computations, the simplest technical description of the analytical aspect of Q-methodology is that it entails the factor analysis of the transpose of the data matrix. In other words, instead of reducing the dimensionality of the variables, it reduces the dimensionality of the observations (in this case, respondents), or it seeks clusters of statistically similar respondents. Our approach is not quite Q-methodology, in that we use the data analysis and interpretation aspects of Q-methodology without fully following the data collection approach prescribed by the approach. We explain the difference in data collection techniques at the end of this section.

Q-methodology is a method for obtaining and analyzing opinions. Thus, it consists of collecting subjective information, which is then analyzed using statistical techniques. Stephenson (1935, 1953), the originator of the technique, referred to it as a tool for correlating persons rather than variables. The source materials for Q-methodology are obtained from a concourse (Stephenson 1978), which, simply put, is a detailed discussion on a topic. The purpose of Q-methodology is to distil the essence of the concourse or to identify its essential structure. Statements describing the subjective opinions expressed in the concourse are called a Q-sample. Participants sort these statements on a scale that indicates total agreement to total disagreement with the opinion expressed in each statement. These statements are given to the participants who conduct a Q-sort and assign values to the statements. Pair-wise correlations, referred to as similarities, between the scores assigned by the individuals are computed, hence the reference to correlation between individuals and not variables. These correlations yield a matrix of similarities of opinions among the participants, which is subsequently used in the factor analysis. The factors obtained from this matrix capture and summarize the subjective opinions of the individuals.

Computationally, the data analysis procedure may be described as follows. Data are generally presented in the form of a $(N \times p)$ matrix of N observations (rows, persons) and p variables (columns). Factor analysis (R-technique) uses the (p×p) matrix of correlations between the variables to obtain a reduced number of h (≤p) factors. A full rank matrix would produce p factors, but the purpose of factor analysis is to identify a subset of h factors that capture the majority of the information contained in the full set of p factors. This reduced set of factors captures most of the information contained in the p variables. So, in our case, the data consist of a Q-sort of the deck of cards. In addition to the sort, we have scores ranging from 1 to 100 assigned to each card by the participants. Thus, our data consist of the subjective opinion or order of preference for each of the cards from each participant. The data analysis task is to determine whether the opinions of the participants can be reduced to one dominant opinion, expressed in terms of one common sort of the cards. However, unless the preferences among the participants are homogeneous, it is rare that there would be just one common preference ordering of the cards. The more likely situation is one in which there are two or more perspectives, each represented by its preferred sorting of the cards. More formally, we follow Comrey and Lee (1992, 21–22) and write,

$$z_{ki} = a_{k1}F_{1i} + a_{k2}F_{2i} + \ldots + a_{kh}F_{hi} + a_{ks}S_{ki} + a_{ke}E_{ki}$$

where

z_{ki} is a standard score for person i on data variable k;
a_{k1} is a factor loading for data variable k on common factor 1;
a_{k2} is a factor loading for data variable k on common factor 2;
a_{kh} is a factor loading for data variable k on the last common factor, h;
a_{ks} is a factor loading for data variable k on specific factor k;
a_{ke} is a factor loading for data variable k on error factor k;
F_{1i} is a standard score for person i on common factor 1;
F_{2i} is a standard score for person i on common factor 2;
F_{km} is a standard score for person i on the last common factor, m;
S_{ki} is a standard score for person i on specific factor k;
E_{ki} is a standard score for person i on error factor k.

The z value is a standardized score obtained from the actual data. The other standardized values, F, S, and E are all obtained through the factor analysis, but are not typically displayed by commonly used statistical software because this information is not generally used in estimating the factor loadings. The usual output of a factor analysis is the factor loadings, a, which range from −1 to +1.

Computationally, Q-methodology utilizes factor analysis as expressed above. However, Q-methodology uses factor analysis of the (N × N) matrix of *similarities* between the observations rather than the matrix of correlations between the variables. Recall that the purpose of this analysis is to ascertain whether there are clusters of like-minded respondents; so we are interested in identifying groups of individuals, hence the focus is on the observations. Standard factor analysis is a technique for reducing the number of different measurements, hence the focus there is on the variables.

The maximum number of factors that can be obtained is min(N, p). Assuming that the observations are individuals, the factors of this matrix are composites of these individuals and are often interpreted as characteristic groups or typical persons or, as in our case, value orientations.

By studying the factor loadings, it is possible to begin to characterize composite individuals in terms of how they load onto each factor. Hence, two individuals who have similar preference structures, even though their mean scores and the variability in the scores are different, will be similar in that they will have high loadings on the same factor. Using these factors, it is possible to obtain the mean scores for the attributes. This information about how each group (factor) values these different attributes can yield useful information on both the structure of group preferences for attributes and differences across groups.

Q-methodology is more than a data analysis tool that employs factor analysis. In fact, an integral part of the method is the elicitation of subjective information from the respondents. The principles of self-reference in Q-methodology require that the data (Q-samples) are to be obtained through written and oral communication with the respondents who then sort (Q-sort) these data. The factor analysis of these Q-sorts clusters the respondents according to their preference structures.

In our study, we obtained the attributes and their values from discussions with the people, some of whom we eventually asked for their preferences. However, there was only partial overlap between the people from whom we obtained the information that served as the basis of the responses and those from whom we sought information regarding preferences. Also, we did not feed the precise language obtained from the respondents back to them. Hence, we lost some of the subjectivity that distinguishes Q-methodology from other preference elicitation and analysis methods. In a strict sense, we are not implementing Q-methodology and therefore refer to our analysis as Q-type analysis.

Data Format

The card-sort exercise yielded seventy usable decks of twenty-three cards each. Each respondent gave a rating between 1 and 100 to each card in his or her deck. The data set thus consisted of 1,610 (23×70) rows where each row represents one of twenty-three cards scored by one of the seventy respondents. The three levels of each attribute were represented by two dichotomous variables indicating the presence or absence of each of the three levels. The twenty-one columns of the data matrix correspond to the score associated with each card (one column) and the ten attributes at three levels each, yielding the remaining twenty columns of (0,1) dichotomous variables. The data value (0 or 1) in each column indicates the level of the attribute represented on the card. Conjoint analysis uses this data matrix to estimate the regression of the scores on the attribute levels to obtain the utility of each attribute level and the importance scores for the attributes.

The Q-type analysis data matrix may be viewed as a transpose of the data matrix used in the conjoint analysis. Except, we first expand the twenty (0,1) coded attribute columns into thirty (0,1) coded columns in which 0's or 1's have the standard meaning representing the absence or presence of the attribute on the card. We premultiply this ($1,610 \times 30$) matrix by a ($1 \times 1,610$) vector of scores to replace the 1's in the matrix by the score given to each card by the seventy respondents. We next transpose this matrix so that the rows represent the thirty attributes and the columns represent the seventy respondents. Hence, we now have a 690×70 matrix where the 690 rows consist of the twenty-three cards representing the presence or absence of the thirty attribute-levels, and the seventy columns now represent the individual respondents.

Q-type analysis entails a factor analysis of the (70×70) variance-covariance (similarity) matrix obtained from this data matrix representing the scores on the thirty attribute levels given by seventy respondents using a deck of twenty-three cards. Instead of the correlation matrix, which is generally used as the basis of a factor analysis to reduce the dimensionality of the set of variables, we are now using this similarity matrix as the basis of the factor analysis used to reduce the dimensionality of the set of respondents. The rank of this 70×70 matrix is twenty-three because all the variance in the data is obtained from the scoring of the twenty-three cards.

Chapter 10

Managing Wicked Environmental Problems

In chapter 1, we briefly mentioned some historical factors that may contribute to the transformation of complicated problems into wicked problems. These include broad, structural socioeconomic and demographic changes and also more immediate triggering events that polarize public debate and shift the political dynamics of environmental management dilemmas. Here we explore these themes in more detail. We also present and discuss three take-home lessons for public managers. We suggest that a public manager facing a wicked problem should (1) stop looking for the perfect solution; (2) seek instead a *satisficing* response; and (3) consider applying the iterative, analytic, adaptive, participatory process described in this book (particularly in chapters 6–9).

In this closing chapter, we take the perspective of the public manager. Despite our emphasis in previous chapters on public participation and stakeholder engagement, the public manager retains a special position. The public manager is the person at the center of the controversy, pulled in various directions by interest groups, activists, local community members, the private sector, the broader public, the courts, the media, elected officials at various levels, and the agency and administration leadership in the state and national capitals. Moreover, the public manager may be the person most likely to recognize that he or she is facing a wicked problem, and thus most likely to be in position to initiate the enhanced *learning network* process that we recommend. Most important, the public manager is the person required by law and job description to make and implement environmental management decisions.

In focusing on the public manager's perspective in this closing chapter, we draw on the experience of Ron Stewart, one of the authors of this book. Ron was regional forester for the Sierra Nevada national forests in the early 1990s. In retrospect, we can see that this was the time when forest planning in the region changed from being a complicated problem to being a wicked problem. As is the case with many public managers immersed in the demands of the moment, however, Ron did not have the opportunity when wrestling with the Sierra Nevada Forest Plan Amendment process to integrate the relevant literature from diverse disciplines and apply it to his highly pressurized situation. Nor did he have time to consider the broad social changes that had shifted the policy ground beneath his feet. Consequently he did not have a name or a big-picture explanation for the new *wickedness* that was making his job so difficult.

After examining the topic with some scholarly detachment over the past several years, however, Ron and the rest of us on the research team have come to what we believe is a richer understanding of the dilemma that he faced in the Sierra Nevada, and that other public managers continue to face in the United States and elsewhere. While we focus on public managers, we believe that the recommendations and insights summarized in this chapter will also be useful for other public and private stakeholders involved in wicked problems and for scholars and students interested in policy processes in these challenging circumstances.

Take-Home Lessons

The first take-home lesson we offer to the public manager facing a wicked problem is to stop looking for the perfect solution. Public environmental managers—who typically have substantial scientific or technical training and expertise—tend to believe there is a *correct* or *best* policy response for any given environmental management dilemma. It follows that the manager's job is to find that optimal response. Even when recognizing that a particular problem may be too complicated to address fully through any one action, public managers still expect that by developing a list of possible policy alternatives and carefully comparing the strengths and weaknesses of each they will be able to identify the best action. If the best action does not emerge from this systematic selection process, they assume that the primary obstacle is insufficient data and analysis. They undertake further studies, expecting that at some point the weight of evidence will make the correct choice both clear to them and convincing to other stakeholders and the general public.

A key lesson from this book, and one that may go against the training and inclinations of many expert public managers, is that some problems—that is, wicked problems—simply do not have correct or best solutions. This is the case for at least three fundamental reasons. First, for this type of policy dilemma, there is no single, broadly accepted definition of the problem. Second, the relevant science is inherently uncertain. And third, the policy-making context is dynamic and unstable—meaning that public values and opinions, the level of budgetary and political support, the administrative leadership, and other key factors are likely to shift over time in unpredictable ways. In such circumstances, collecting and analyzing more scientific or ecological data will not bring a solution any closer. Instead, short-term political processes and outcomes become centrally important.

With this lesson in mind, the public manager should put aside the idea of finding an optimal solution and turn instead—following the second of our three take-home lessons—to the messy, confounding, frustrating, but ultimately unavoidable task of seeking a satisficing response.

By definition, an optimal solution claims to be both a solution to the problem and the best solution among the various alternatives available. Neither of these characteristics can be attained in the wicked problem context. A satisficing response, in contrast, makes no claim to be an optimal solution. Rather it is a response that acknowledges and attempts to work within the exceptional challenges presented by a wicked problem. It aims at the more limited goals of easing the gridlock and laying a foundation for progress over time. A satisficing response is necessarily adaptive and inclusive. While it cannot hope for universal acceptance, it may gradually win wider support than other options as it contributes to diminishing polarization.

To meet the requirements of the public manager in practice, however, the satisficing response must still meet certain minimal standards of adequacy. It must be ecologically feasible, satisfy current laws and regulations, fit realistically within budget constraints, and show some promise of contributing to group learning and trust building. We emphasize that in the wicked problem context, however, even these modest goals will be extremely difficult to achieve. Generating a satisficing response will require the technically trained public manager to venture deep into the social and political thickets of stakeholder engagement, compromise, and mutual learning. Moreover, a satisficing response is not a goal that will be achieved at a single point on a policy-making timeline. Establishing and maintaining a satisficing response will inevitably be a long-term, iterative, and adaptive project.

This leads to our third take-home lesson: The public manager facing a wicked problem should consider the enhanced learning network process that we outline in this book. In principle, the initiative for implementing the process could come from any of the key stakeholders involved in the dilemma. But as the focal point of the pressures, the public manager is likely to be well placed to introduce this option.

While our recommended process cannot ensure success, we believe it holds promise as a way to begin to move a wicked problem from a state of intractability toward a satisficing response. As discussed in chapter 7, the proposed process rests on two foundational components. The first component is a version of the iterative, analytic, deliberative, adaptive, and participatory learning network approach described by a committee of the US National Research Council (1996). As we discuss throughout this book, the multidisciplinary literature relevant to wicked problems strongly suggests that progress requires effective participation of key stakeholders in a process that builds mutual learning and trust through repeated interactions over time informed by the best available science. We emphasize that the term *stakeholders* in this context should be interpreted broadly to include public agency personnel, technical experts inside and outside the public sector, and representatives of groups reflecting an inclusive spectrum of the attitudes, values, and priorities of the broader public.

The second essential component of our recommended approach is the NEPA process—the decision-making procedures that emerged from the requirements of the National Environmental Policy Act (NEPA) of 1969. These requirements include detailed environmental impact statements and multiple opportunities for public input. Any process aiming to address wicked environmental problems must allow the public manager to meet these statutory and regulatory requirements. We keep the NEPA guidelines in mind in developing our proposal in chapters 6 and 7 because these rules, or similar sets of rules influenced by NEPA, now form a widely accepted standard across the United States and in many countries worldwide. Thus these guidelines can serve as a generic regulatory framework within which our proposed process can be implemented. However, public managers can adapt the process to other national or local requirements that they may have to meet. In describing our recommended process, we illustrate how the learning network approach recommended by the National Research Council (1996) can be integrated with the NEPA process (see fig. 6.2).

On this foundation, we add innovative steps designed to strengthen both the social and scientific components of the combined process. We suggest that an enhanced learning network process should include formal, quantitative elicitation and analysis of stakeholder preferences. Participatory processes in current use generally provide multiple opportunities for public input. But the opinions and priorities expressed through these forums may not accurately capture participants' underlying beliefs. Participants, for example, may articulate relatively extreme positions as negotiating ploys, or they may frame their arguments in rational or scientific terms when their positions are in fact primarily value based. And these behaviors are common to technical experts as well as to members of interest groups and the general public.

We argue that surveys and other preference elicitation exercises used in marketing, political polling, and related activities can usefully be applied to wicked problems. Data collected through these techniques may then be analyzed to reveal overlooked or underappreciated aspects of stakeholder attitudes. In chapters 8 and 9, we describe examples of these preference-elicitation techniques, along with the quantitative tools that can be used to analyze the resulting data, and present the results of a pilot study using these methods that we conducted in the Sierra Nevada.

These steps of preference elicitation and data analysis are of more than scholarly interest. They should be built in to the learning network process. Participants in learning network and decision-making processes for environmental management need access to sophisticated analyses of both the best available ecological data and the best available attitudinal data. As participants work through their iterative deliberations, detailed analytic information on stakeholder preferences may help them find opportunities for compromise and progress that would otherwise be missed. Moreover, as group learning takes place, participant preferences and values may shift. Continuing elicitation and analysis of stakeholder attitudes may capture these shifts and contribute usefully to ongoing deliberations.

We argue further that data from surveys and other preference-elicitation exercises can be used to support statistical simulations of how stakeholders may respond over time to ecological outcomes predicted to follow from various policy alternatives under consideration. As we describe in chapter 9, the integration of attitudinal models and ecological models may allow stakeholders to rank their responses to alternative projected scenarios more effectively. Findings from such simulations have the potential to reduce the conflict and polarization that tend to characterize the public debate of policy options reflecting deeply held differences in priorities and values.

Furthermore, as discussed in chapter 5, the enhanced learning network process we envision will almost certainly lead to policies that depend on adaptive management. We believe that including improved analysis of stakeholder preferences in the decision-making process will support more effective application of adaptive management to both the ecological and social dimensions of wicked environmental problems.

Wickedness in a Changing Society

For the public manager facing a wicked problem, the act of identifying the problem as wicked is an essential first step. This recognition in itself will not move the problem closer to a solution, but it may help participants understand the dilemma's intransigence and open up alternative responses. It should lead the public manager to act on the first take-home lesson mentioned above—that is, to stop looking for the perfect solution. The manager and other stakeholders may then be able act on the other two take-home lessons as well—that is, to seek a satisficing response through a learning network process.

In discussing the take-home lessons, we mentioned that from the public manager's perspective a satisficing solution should contribute to trust building. This raises the question of why trust in public agencies is lacking in many environmental controversies. In this section, we describe broad-based social changes occurring over the past several decades that have contributed significantly to the more common emergence of wickedness in environmental management. These trends have contributed to wickedness by creating conditions that foster conflicts between the agencies and various sectors of the public, which, in turn, have undermined trust on both sides. Keeping in mind the four case studies introduced in chapter 3, we briefly describe these trends as they were experienced in the United States, Europe, and Tanzania. Understanding these trends may help public managers and other stakeholders recognize when they are enmeshed in a wicked problem.

Public agencies with responsibilities for managing natural resources typically have high levels of technical competence. Historically, other branches of government and the broader public have generally held these agencies in relatively high esteem and have afforded them considerable latitude to fulfill their responsibilities without micromanaging from a distance. For the first three-quarters of the twentieth century, for example, the general public in the United States largely deferred to the Forest Service, the Army Corps of Engineers, and other relevant federal agencies, trusting them to manage natural resources in the national interest.

Political scientists, however, point to structural social and economic changes beginning in the 1970s and 1980s in the United States that shifted public attitudes regarding the environment and undermined the public's trust in environmental management agencies (Kraft and Vig 2010). During early and middle periods of industrialization, societies are generally willing to tolerate significant levels of environmental degradation in return for rapid economic growth. This pattern was reflected in environmental management in the United States up through the 1970s. As we discussed in reviewing the South Florida case in chapter 3, for example, the Army Corps of Engineers in the postwar decades had broad social and political support as it managed watersheds to enhance navigation, open access to new land, and provide water for agricultural, commercial, and residential development. Adverse impacts on wildlife and habitat and other environmental amenities were largely ignored. Similarly, the Forest Service managed the vast public lands under its jurisdiction primarily to provide timber, grazing range, and other valuable natural resources to the growing economy, again with a broad social consensus supporting these policies.

In later stages of industrialization, however, trends shift. With the emergence of a broad-based middle class with basic material needs met, people begin to pay more attention to environmental degradation. With the attendant growth in wealth and income, industrialized societies also find that they can afford the economic costs of more stringent environmental policies. Consequently, priorities shift. The public becomes more concerned about environmental issues, and the environmental impact of economic activities comes under heightened scrutiny. The environment then rises on political agendas, and policy makers institute regulations to mitigate environmental harms. This pattern of environmental degradation initially being accepted during industrialization but later becoming a public concern as average incomes increase has been observed across many societies as they develop (Kraft and Vig 2010).

The United States appears to have reached the level of economic development that brought it to the tipping point of changing attitudes on the environment in the late 1960s. As public attitudes shifted, the political system responded. The National Environmental Policy Act was the first in a wave of major US environmental laws enacted in the 1970s. Along with regulating activities that harmed the environment, these laws also typically included explicit provisions requiring public input and permitting citizen lawsuits.

Broadly analogous social and political changes also occurred in Europe during the same period (Judt 2005). Given the diversity of economic, cultural, and historical circumstances across Europe, however, there was wide variation in the timing, strength, and form of these shifts. In northern and western European countries, the timing matched relatively closely to that seen in the United States. The process occurred later and with different political dynamics in Mediterranean and eastern Europe. Spain, Portugal, and Greece, for example, remained under various forms of dictatorship until the 1970s, and Eastern European countries were behind the Iron Curtain until 1989. Autocratic or totalitarian government policies in these countries severely retarded economic development in comparison with the rest of the continent. These historical patterns help explain why northern and western European countries have been the primary drivers of the European Union's efforts to regulate greenhouse gas emissions and why the poorer countries of the former Soviet bloc tend to be less enthusiastic in their support.

Although Tanzania has not yet come close to reaching the level of per capita income associated with rising environmental consciousness in the United States and western Europe in the 1960s and 1970s, changes in environmental attitudes in developed countries have had an impact in developing countries as well. Following the Earth Summit in Rio de Janeiro in 1992, rich counties and multilateral institutions such as the United Nations and the World Bank began to tie a portion of their foreign aid directly to environmental programs in beneficiary nations. Donor demands and incentives led many developing countries to set aside national parks for the protection of wildlife and habitat. At the same time, international nature tourism expanded rapidly. Developing countries with environmental attractions—wildlife in Tanzania, for example—saw increased foreign investment in tourism facilities and increased foreign exchange earnings from the tourists themselves. This in turn encouraged further conservation efforts to protect and expand the new revenue streams.

In both rich and poor countries, increasingly rapid demographic changes linked to the economic and attitudinal shifts described here brought environmental management agencies into conflict with various sectors of the public. In the United States, a growing population with growing levels of education and wealth meant that more people, often with differing priorities and values, came into contact with the lands and waters managed by federal agencies. In trends that began slowly but accelerated in the last quarter of the twentieth century, rural communities with livelihoods based directly on local natural resources were reshaped by summer visitors and resident commuters who sought recreational opportunities, lower land values, and bucolic settings. Agencies such as the Forest Service, which in the past had operated with little public scrutiny or dissent, began to experience growing political pressures to change their management objectives and practices.

At the same time, new environmental laws requiring public participation and permitting citizen lawsuits contributed to rising contentiousness and litigiousness. Interest groups from across the spectrum, including environmental organizations and trade associations, competed with increasing sophistication and intensity to push policies in the directions that their members favored. Revolutions in telecommunications and information technology further accelerated these trends by freeing interest groups from geographical limits as they recruited active participants. In many environmental management controversies, *communities of interest* have come to exceed in influence the *communities of place* that traditionally interacted locally with the Forest Service and other public environmental management agencies.

In a third trend complicating environmental policy making, scientists and managers have come to recognize over the past several decades the importance of planning at the scale of ecosystems or landscapes, rather than at the typically smaller scale of arbitrary political jurisdictions. As the reach of environmental planning expands, so do the number and diversity of stakeholder groups and the potential for conflict.

With somewhat different social dynamics, demographic changes are also increasing conflicts for environmental management agencies in developing countries. As discussed in chapter 3, for example, growing numbers of local people and international visitors in and near Ngorongoro Crater in Tanzania are threatening the wildlife and habitat that make the area both an important tourist destination and a significant source of income for the nation. But if the public management authority attempts to limit the number of tourists, it comes into conflict with the tourism industry, and often with the central government administration as well. Conversely, if it acts to limit the activities of local communities, it becomes a target of resentment and anger from this direction. The Maasai commonly accuse the agency of implementing policies that favor wildlife over local people. In the 1950s and 1960s, Tanzanian government agencies could unilaterally relocate local residents in order to establish or protect nature reserves with minimal criticism from beyond the affected communities. Given current international norms, however, such actions are no longer politically feasible.

Thus significant socioeconomic and demographic changes occurring over a period of decades have undermined the default social consensus that public environmental managers could count on for support. More stakeholders are now involved in almost every decision, and these stakeholders are more diverse in their values and priorities and have far greater access to legal and financial resources. At the same time public managers, following new scientific understanding, aim to address environmental concerns at larger scales. Consequently, conflicts have become more common, participants have become more politicized, and the level of trust has declined. Once trust is lost, it is difficult to regain. Bureaucracies are slow to change, and in conditions of distrust even delayed positive changes may be viewed with suspicion. With these broad trends operating in the background, the emergence of specific conflicting policy priorities within a given management context can trigger wickedness. This happened in all the cases we describe in chapter 3. As we suggest in chapters 6–9, a learning network process, combined with enhanced elicitation and analysis of stakeholder attitudes, may have the potential to rebuild lost trust and thus moderate entrenched wickedness.

Conclusions

In the Sierra Nevada, key issues that divide stakeholders include protecting spotted owls and old-growth habitat, harvesting timber, promoting recreational activities, and reducing wildfires. In South Florida, conflicting priorities include restoring the Everglades while managing floods and providing sufficient water for the region's burgeoning cities. In the European Union's program to reduce greenhouse gas emissions, competing concerns include reducing potential threats from future climate change while taking into account income inequality across the European Union and maintaining satisfactory economic growth. In Tanzania, conflicting goals include conserving wildlife and mitigating the adverse effects of tourism while maintaining the revenue stream and addressing local poverty. Addressing any one component of any of these controversies would be a complicated problem. Attempting to address all the components simultaneously—in the face of limited budgets, scientific and administrative uncertainty, polarized public values, and lack of trust in the managing agency—is a wicked problem.

For public managers embroiled in particular seemingly intractable dilemmas, the underlying social and demographic trends contributing to a problem's wickedness may seem distant and of limited relevance. But a fuller understanding of the nature of wicked problems, both their roots and identifying characteristics, may help public managers and other key stakeholders recognize the constraints that wickedness imposes upon them and work together more effectively within those limitations. As we have discussed throughout the book, useful steps along this path may include abandoning the search for optimal solutions and seeking instead progress toward satisficing outcomes through the development of enhanced learning networks.

REFERENCES

Allan, C., and A. Curtis. 2005. "Nipped in the Bud: Why Regional Scale Adaptive Management is Not Blooming." *Environmental Management* 36 (3): 25–41.
Allen, G. M., and E. J. Gould Jr. 1986. "Complexity, Wickedness, and Public Forests." *Journal of Forestry* 84 (4): 20–23.
Arstein, S. R. 1969. "A Ladder of Citizen Participation." *Journal of American Institute of Planners* (35): 216–24.
Arvai, J., R. Gregory, D. Ohlson, B. Blackwell, and R. Gray. 2006. "Letdowns, Wake-up Calls, and Constructed Preferences: People's Responses to Fuel and Wildfire Risks." *Journal of Forestry* 104 (4): 173–81.
Asher, W. 2001. "Coping with Complexity and Organizational Interests in Natural Resource Management." *Ecosystems* (4): 742–57.
Auberson-Huang, L. 2002. "The Dialogue Between Precaution and Risk." *Nature Biotechnology* (20): 1076–78.
Balint, Peter J. 2006. "Improving Community-Based Conservation near Protected Areas: The Importance of Development Variables." *Environmental Management* 38 (1): 137–48.
Barbour, Michael G., and Jack Majors, eds. 1990. *Terrestrial Vegetation of California*. Sacramento: California Native Plant Society.
Bardach, E. 1996. *The Eight-Step Path of Policy Analysis: A Handbook for Practice*. Berkeley, CA: Berkeley Academic Press.
Baucum, L. E., and R. W. Rice. 2009. *An Overview of Florida Sugarcane*. SS-AGR-232. Gainesville: University of Florida Institute of Food and Agricultural Sciences. http://edis.ifas.ufl.edu/pdffiles/SC/SC03200.pdf.
BBC News. 2004. "Global Warming 'Biggest Threat.'" January 9. http://news.bbc.co.uk/1/hi/3381425.stm.
Beck, U. 1992. *Risk Society: Toward a New Modernity*. London: Sage.
Beierle, Thomas C. 2002. "The Quality of Stakeholder-based Decisions." *Risk Analysis* 22 (4): 739–49.
Beierle, Thomas C., and Jerry Cayford. 2002. *Democracy in Practice: Public Participation in Environmental Decisions*. Washington, DC: Resources for the Future.
Beierle, Thomas C., and David M. Konisky. 1999. "Public Participation in Environmental Planning in the Great Lakes Region." In *Discussion Paper 99-50*. Washington, DC: Resources for the Future.

Bhat, Mahadev, and Athena Stamatiades. 2003. "Institutions, Incentives, and Resource Use Conflicts: The Case of Biscayne Bay, Florida." *Population and Environment* 24 (6): 485–509.

Block, J. 1961. *The Q-sort Method in Personality Assessment and Psychiatric Research*. Springfield, IL: Charles C. Thomas (reprinted in 1978 by Consulting Psychologists Press, Palo Alto, CA).

Bodansky, D. 1991. "Scientific Uncertainty and the Precautionary Principle." *Environment* 33 (7): 4–5, 43–44.

Boko, M., I. Niang, A. Nyong, C. Vogel, A. Githeko, M. Medany, B. Osman-Elasha, R. Tabo, and P. Yanda. 2007. "Africa." In *Climate Change 2007: Impacts, Adaptation and Vulnerability. Contribution of Working Group II to the Fourth Assessment Report of the Intergovernmental Panel on Climate Change*, edited by M. L. Parry, O. F. Canziani, J. P. Palutikof, P. J van der Linden, and C. E. Hanson, 433–67. Cambridge: Cambridge University Press.

Bormann, B. T., P. G. Cunningham, M. H. Brookes, V. W. Manning, and M. W. Collopy. 1994. *Adaptive Ecosystem Management in the Pacific Northwest*. PNW-GTR-341. Portland, OR: USDA Forest Service Pacific Northwest Research Station. http://www.treesearch.fs.fed.us/pubs/8965.

Bradbury, B., W. Nehlsen, T. E. Nickelson, K. M. S. Moore, R. M. Hughes, D. Heller, J. Nicholas, D. L. Bottom, W. E. Weaver, and R. L. Beschta. 1995. *Handbook for Prioritizing Watershed Protection and Restoration to Aid Recovery of Native Salmon*. Eugene, OR: Pacific Rivers Council. http://www.salmonhabitat.org/library/bradbury_et_al.pdf.

Brams, Steven J., and M. Remzi Sanver. 2009. "Voting Systems That Combine Approval and Preference." In *The Mathematics of Preference, Choice and Order*, edited by S. J. Brams, W. V. Gehrlein, and F. S. Roberts. Berlin: Springer-Verlag.

Breyer, S. 1993. *Breaking the Vicious Circle: Toward Effective Risk Regulation*. Cambridge, MA: Harvard University Press.

Brown, P. 1992. "Popular Epidemiology and Toxic Waste Contamination: Lay and Professional Ways of Knowing." *Journal of Health and Social Behavior* 33: 267–81.

Brown, S. R., W. Durning, and S. Selden. 1999. "Q-methodology." In *Handbook of Research Methods in Public Administration*, edited by Gerald J. Miller and Marcia L. Whicker. New York: M. Dekker.

Buck, Louise E., Charles C. Geisler, John Schelhas, and Eva Wallenberg, eds. 2001. *Biological Diversity: Balancing Interests through Adaptive Collaborative Management*. Washington, DC: CRC Press.

Burns, T. R., and T. Dietz. 1992. "Technology, Sociotechnical Systems, Technological Development: An Evolutionary Perspective." In *New Technology at the Outset: Social Forces in the Shaping of Technological Innovation*, edited by M. Dierkes and U. Hoffman. Frankfurt: Campus Verlag.

Butler, K. F., and T. M. Koontz. 2005. "Theory into Practice: Implementing Ecosystem Management Objectives in the USDA Forest Service." *Environmental Management* 35 (2): 138–50.

Caldwell, L. K. 1998. *The National Environmental Policy Act: An Agenda for the Future*. Bloomington, IN: Indiana University Press.
Cameron, J., and J. Aboucher. 1991. "The Precautionary Principle: A Fundamental Principle of Law and Policy for the Protection of the Global Environment." *Boston College International and Comparative Law Review* 14: 1–27.
Carroll, Matthew S., Keith A. Blatner, Patricia J. Cohn, and Todd Morgan. 2007. "Managing Fire Danger in the Forests of the US Inland Northwest: A Classic 'Wicked Problem' in Public Land Policy." *Journal of Forestry* 105 (5): 239–44.
Cave, Damien. 2008. "Florida Buying Big Sugar Tract for Everglades." *New York Times*, June 25. http://www.nytimes.com/2008/06/25/us/25everglades.html?_r=1.
Checkland, P. 2005. "Webs of Significance: The Work of Geoffrey Vickers." *Systems Research and Behavioral Science* 22 (4): 285–90.
Checkland, P., and A. Casar. 1986. "Vickers' Concept of an Appreciative System: A Systemic Account." *Journal of Applied Systems Analysis* 13: 3–17.
Checkland, P. B., and J. Poulter. 2006. *Learning for Action: A Short Definitive Account of Soft Systems Methodology and Its Use for Practitioners, Teachers and Students*. Chichester, UK: Wiley.
Chilvers, Jason. 2008. "Deliberating Competence: Theoretical and Practitioner Perspectives on Effective Participatory Appraisal Practice." *Science, Technology & Human Values* 33 (2): 155–85.
Churchman, C. W. 1967. "Wicked Problems." *Management Science* 14 (4): B141–B142.
Clarke, Alice, and George Dalrymple. 2003. "$7.8 Billion for Everglades Restoration: Why Do Environmentalists Look So Worried?" *Population and Environment* 24 (6): 541–69.
Cohen, B. L. 1987. "Reducing the Hazards of Nuclear Power: Insanity in Action." *Physics and Society* 16: 2–4.
Comba, P., M. Martuzzi, and C. Botti. 2002. "Interpreting the Precautionary Principle: A Comparison between a Bayesian Approach and One Based on the Maximin Principle." *Epidemiology* 13 (4): 56.
Combs, B., and P. Slovic. 1979. "Newspaper Coverage of Causes of Death." *Journalism Quarterly* 56: 837–43, 849.
Committee of Scientists. 1999. *Sustaining the People's Lands: Recommendations for Stewardship of the National Forests and Grasslands into the Next Century*. Washington, DC: US Department of Agriculture. http://www.fs.fed.us/news/news_archived/science/cosfrnt.pdf.
Comrey, A., and H. B. Lee. 1992. *A First Course in Factor Analysis*, 2nd ed. Hillsdale, NJ: Lawrence Erlbaum Associates.
Conklin, E. J. 2006. *Dialogue Mapping: Building Shared Understanding of Wicked Problems*. Chichester, UK: Wiley.

Conklin, E. J., and W. Weil. 1997. Wicked Problems: Naming Pain in Organizations. Group Decision Support Systems. http://www.leanconstruction.org/pdf/wicked.pdf.

Cooke, Bill, and Uma Kothari, eds. 2001. *Participation: The New Tyranny?* London: Zed Books.

Cooney, R., and B. Dickson, eds. 2005a. *Biodiversity and the Precautionary Principle: Risk and Uncertainty in Conservation and Sustainable Use.* London: Earthscan.

———. 2005b. "Precautionary Principle, Precautionary Practice: Lessons and Insights." In *Biodiversity and the Precautionary Principle: Risk and Uncertainty in Conservation and Sustainable Use*, edited by R. Cooney and B. Dickson, 287–98. London: Earthscan.

Copenhagen Consensus. 2008. *The Outcome of the Copenhagen Consensus 2008.* http://www.copenhagenconsensus.com/Home.aspx.

Council on Environmental Policy. 1987. *Memorandum to Agencies: Forty Most Asked Questions Concerning CEQ's National Environmental Policy Act Regulations.* CFR parts 1500–1508. http://ceq.hss.doe.gov/nepa/regs/40/40p1.htm.

———. 2007. *A Citizen's Guide to the NEPA: Having Your Voice Heard.* Washington, DC: Executive Office of the President of the United States. http://ceq.hss.doe.gov/publications/citizens_guide_to_nepa.html.

Covello, V. T., P. M. Sandman, and P. Slovic. 1989. "Risk Communication, Risk Statistics, and Risk Comparisons: A Manual for Plant Managers." In *Effective Risk Communication: The Role and Responsibility of Government and Nongovernment Organizations*, edited by V. T. Covello, D. B. McCallum, and M. T. Pavlova. New York: Plenum Press.

Daly, H. E. 1993. "The Steady-state Economy: Toward a Political Economy of Biophysical Equilibrium and Moral Growth." In *Valuing the Earth: Economics, Ecology, Ethics*, edited by H. E. Daly and K. N. Townsend. Cambridge, MA: MIT Press.

Daniels, S. E., and G. B. Walker. 2001. *Working through Environmental Conflict: The Collaborative Approach.* Westport, CT: Praeger.

Davis, F. W., and D. M. Stoms. 1996. "Sierran Vegetation: A Gap Analysis." In *Sierra Nevada Ecosystem Project: Final report to Congress, vol. 2. Assessments and Scientific Basis for Management Options.* Davis: University of California Centers for Water and Wildland Resources. http://ceres.ca.gov/snep/pubs/web/PDF/VII_C23.PDF.

deFur, Peter L., and Michelle Kaszuba. 2002. "Implementing the Precautionary Principle." *The Science of the Total Environment* 288: 155–65.

Delbeq, A., and A. Van de Ven. 1971. "A Group Process Model for Problem Identification and Program Planning." *Journal of Applied Behavioural Science* 7: 467–92.

DeLeon, Peter. 1995. "Democratic Values and the Policy Sciences." *American Journal of Political Science* 39 (4): 886–905.

Dethlefsen, V. 1993. "The Precautionary Principle: Toward Anticipatory Environmental Management." In *Clean Production Strategies: Developing Preventive*

Strategies in the Industrial Economy, edited by T. Jackson. Washington, DC: CRC Press.

Dietz, T. 1987. "Theory and Method in Social Impact Assessment." *Sociological Inquiry* 57: 54–69.

Dietz, T., S. Frey, and E. Rosa. 2001. "Risk Assessment and Management." In *The Environment and Society Reader*, edited by S. Frey, 272–99. Needham Heights, MA: Allyn & Bacon.

Dietz, T., E. Ostrom, and P. Stern. 2003. "The Struggle to Govern the Commons." *Science* 302 (5652): 1907–12.

Dietz, T., P. Stern, and R. W. Rycroft. 1989. "Definitions of Conflict and the Legitimation of Resources: The Case of Environmental Risk." *Sociological Forum* (4): 47–70.

Douglas, M. 1985. *Risk Acceptability According to the Social Sciences*. New York: Russell Sage Foundation.

Douglas, M., and A. Wildavsky. 1982. *Risk and Culture: The Selection of Technological and Environmental Dangers*. Berkeley: University of California Press.

Dryzek, John S. 2000. *Deliberative Democracy and Beyond*. New York: Oxford University Press.

Dryzek, J. S., and J. Berejikian. 1993. "Reconstructive Democratic Theory." *American Political Science Review* 87 (1): 48–60.

Dryzek, John S., and Christian List. 2003. "Social Choice Theory and Deliberative Democracy: A Reconciliation." *British Journal of Political Science* 33 (1): 1–28.

Duane, T. P. 1996. "Human Settlement, 1850–2040." In *Sierra Nevada Ecosystem Project: Final report to Congress, vol. 2, Assessments and Scientific Basis for Management Options*. Davis: University of California Centers for Water and Wildland Resources. http://ceres.ca.gov/snep/pubs/web/PDF/VII_C11.PDF.

Dunn, W. N. 1994. *Public Policy Analysis: An Introduction*, 2nd ed. Englewood Cliffs, NJ: Prentice Hall.

Durning, Dan. 1993. "Participatory Policy Analysis in a Social-Service Agency: A Case-Study." *Journal of Policy Analysis and Management* 12 (2): 297–322.

Durning, D. 1999. "The Transition from Traditional to Postpositivist Policy Analysis: A Role for Q-methodology." *Journal of Policy Analysis and Management* 8: 389–410.

Durning, D., and D. Edwards. 1992. "The Attitudes of Consolidation Elites: An Empirical Assessment of Their Views of City-County Mergers." *Southeastern Political Review* 20: 355–83.

Egelko, B. 2008. "Court Blocks Bush's Plan for Logging in Sierra." *San Francisco Chronicle*, May 15.

Ellerman, A. Danny, and Paul L. Joskow. 2008. *The European Union's Emissions Trading System in Perspective*. Arlington, VA: Pew Center on Global Climate Change.

Ellerman, A. Danny, Paul L. Joskow, and David Harrison. 2003. *Emissions Trading in the U.S.: Experience, Lessons, and Considerations for Greenhouse Gases*. Arlington, VA: Pew Center on Global Climate Change.

European Commission. 2008. *Special Eurobarometer 295: Attitudes of European Citizens Towards the Environment*. Brussels: Eurobarometer. http://ec.europa.eu/public_opinion/archives/ebs/ebs_295_en.pdf.

Farquhar, Francis P. 1965. *History of the Sierra Nevada*. Berkeley: University of California Press.

Fiorino, Daniel J. 1990. "Citizen Participation and Environmental Risk: A Survey of Institutional Mechanisms." *Science, Technology, & Human Values* 15 (2): 226–43.

———. 2000. "Innovation in U.S. Environmental Policy." *American Behavioral Scientist* 44 (4): 538–47.

Fischer, Frank. 1993. "Citizen Participation and the Democratization of Policy Expertise: From Theoretical Inquiry to Practical Cases." *Policy Sciences* 26 (3): 165–87.

Fischhoff, B. 1990. "Psychology and Public Policy: Tool or Tool Maker?" *American Psychologist* 45: 647–53.

Fischhoff, B., S. Lichtenstein, P. Slovic, S. Derby, and R. Keeny. 1981. *Acceptable Risk*. New York: Cambridge University Press.

Fishbein, M. 1967. "A Behavior Theory Approach to the Between Beliefs about an Object and the Attitude toward the Object." In *Readings in Attitude Theory and Measurement*, edited by M. Fishbein, 389–99. New York: Wiley.

Fisher, A., and Y. Chen. 1996. "Customer Perceptions of Agency Risk Communication." *Risk Analysis* 16 (2): 177–84.

Fone, M., and P. C. Young. 2000. *Public Sector Risk Management*. Oxford: Butterworth-Heinemann.

Freeman, A. Myrick. 2006. "Economic Incentives and Environmental Policy." In *Environmental Policy: New Directions for the Twenty-First Century*, 5th ed., edited by Norman J. Vig and Michael E. Kraft, 193–214. Washington, DC: CQ Press.

Freudenburg, W. R. 1988. "Perceived Risk, Real Risk: Social Science and the Art of Probabilistic Risk Assessment." *Science* 242: 44–49.

Gen, S. 2001. "Progress and Challenges of Adaptive Ecosystem Management at Military Installations: A Case Study of Fort Huachuca, Arizona." *Environmental Practice* 3 (3): 153–62.

Gericke, Kevin L., Jay Sullivan, and J. Douglas Wellman. 1992. "Public-Participation in National Forest Planning: Perspectives, Procedures, and Costs." *Journal of Forestry* 90 (2): 35–38.

Giddens, A. 1990. *The Consequences of Modernity*. Stanford: Stanford University Press.

Gilmour, A., G. Walkerden, and J. Scandol. 1999. "Adaptive Management of the Water Cycle on the Urban Fringe: Three Australian Case Studies." *Conservation Ecology* 3 (1): 11. http://www.ecologyandsociety.org/vol3/iss1/art11/.

Goklany, I. M. 2001. The Precautionary Principle: A Critical Appraisal of Environmental Risk Assessment. Washington, DC: Cato Institute.

Goldman, M. 2006. Sharing Pastures, Building Dialogues: Maasai and Wildlife Conservation in Northern Tanzania. PhD diss., University of Wisconsin–Madison.

Goldstein, Gregg. 2005. "The Legal System and Wildlife Conservation: History and the Law's Effect on Indigenous People and Community Conservation in Tanzania." *Georgetown International Environmental Law Review* 17 (3): 481–514.
Goodnough, Abby. 2007. "Effort to Save Everglades Falters as Funds Drop." *New York Times*, November 2. http://www.nytimes.com/2007/11/02/us/02everglades.html?scp=1&sq=Effort%20to%20save%20Everglades%20falters&st=cse.
Gordan, W. J. J. 1961. *Synectics: The Development of Creative Capacity*. London: Collier-Macmillan.
Gray, B. 1985. "Conditions Facilitating Interorganizational Collaboration." *Human Relations* 38 (10): 911–36.
———. 1989. *Collaborating*. San Francisco: Jossey-Bass.
Green, P. E., and V. Srinivasan. 1978. "Conjoint Analysis in Consumer Research: Issues and Outlook." *Journal of Consumer Research* 5: 102–23.
———. 1990. "Conjoint Analysis in Marketing: New Developments with Implications for Research and Practice. *Journal of Marketing* 54 (4): 3–19.
Greenhalgh, L., and S. A. Neslin. 1981. "Conjoint Analysis of Negotiator Preferences." *Journal of Conflict Resolution* 25 (2): 301–27.
Gregory, Robin, and Katharine Wellman. 2001. "Bringing Stakeholder Values into Environmental Policy Choices: A Community-Based Estuary Case Study." *Ecological Economics* 3: 37–52.
Grin, J., and R. Hoppe. 2000. "Cultural Bias and Framing Wicked Problems." In *Government Institutions: Effects, Changes and Normative Foundations*, edited by H. Wagenaar. Dordrecht: Kluwer Academic Publishers.
Grunwald, Michael. 2002. "Among Environmentalists, the Great Divide." *Washington Post*, June 26.
———. 2006. *The Swamp: The Everglades, Florida, and the Politics of Paradise*. New York: Simon and Schuster.
Gunderson, L. H. 1999. "Resilience, Flexibility and Adaptive Management: Antidotes for Spurious Certitude?" *Conservation Ecology* 3 (1): 7. http://www.ecologyandsociety.org/vol3/iss1/art7/.
———. 2001. "Managing Surprising Ecosystems in Southern Florida." *Ecological Economics* 37 (3): 371–78.
Gunderson, L. H., C. S. Holling, and S. S. Light. 1995. *Barriers and Bridges to the Renewal of Ecosystems and Institutions*. New York: Columbia University Press.
Habron, G. 2003. "Role of Adaptive Management for Watershed Councils." *Environmental Management* 31 (1): 29–41.
Haight, D., and C. Ginger. 2000. "Trust and Understanding in Participatory Policy Analysis: The Case of the Vermont Forest Resources Advisory Council." *Policy Studies Journal* 28 (4): 739–59.
Hammond, John S., Ralph L. Keeney, and Howard Raiffa. 1999. *Smart Choices: A Practical Guide to Making Better Life Decisions*. Boston: Harvard Business School Press.

Holling, C. S., ed. 1978. *Adaptive Environmental Assessment and Management*. London: Wiley.

Homewood, K., E. F. Lambin, E. Coast, A. Kariuki, I. Kikula, J. Kivelia, M. Said, S. Serneels, and M. Thompson. 2001. "Long-Term Changes in Serengeti-Mara Wildebeest and Land Cover: Pastoralism, Population, or Policies?" *Proceedings of the National Academy of Sciences of the United States of America* 98 (22): 12544–49.

Homewood, K. M., and W. A. Rodgers. 2004. *Maasailand Ecology: Pastoralist Development and Wildlife Conservation in Ngorongoro, Tanzania*. Cambridge: Cambridge University Press.

Hough, F. B. 1878. *Report Upon Forestry*. Washington, DC: US Government Printing Office.

Hull, Kathleen L. 2007. "The Sierra Nevada: Archaeology in the Range of Light." In *California Prehistory: Colonization, Culture, and Complexity*, edited by Terry L. Jones and Kathryn L. Klar, 177–90. Lanham, MD: AltaMira.

Interagency SEIS Team. 1994. *Record of Decision for Amendments to Forest Service and Bureau of Land Management Planning Documents within the range of the Northern Spotted Owl: Standards and Guidelines for Management of Habitat for Late-Successional and Old-Growth Forest Related Species within the Range of the Northern Spotted Owl*. Portland, OR: Interagency SEIS Team. http://www.reo.gov/library/reports/newroda.pdf.

Irvine, K. N., and S. Kaplan. 2001. "Coping with Change: The Small Experiment as a Strategic Approach to Environmental Sustainability." *Environmental Management* 28 (6): 713–25.

Johnson, F., and K. Williams. 1999. "Protocol and Practice in the Adaptive Management of Waterfowl Harvests." *Conservation Ecology* 3 (1): 8. http://www.ecologyandsociety.org/vol3/iss1/art8/.

Johnson, Renee J., and Michael J. Scicchitano. 2000. "Uncertainty, Risk, Trust, and Information: Public Perceptions of Environmental Issues and Willingness to Take Action." *Policy Studies Journal* 28 (3): 633–47.

Jordan, A., and T. O'Riordan. 1999. "The Precautionary Principle in Contemporary Environmental Policy and Politics." In *Protecting Public Health and the Environment: Implementing the Precautionary Principle*, edited by C. Raffensperger and J. Tickner. Washington, DC: Island Press.

Judt, Tony. 2005. *Postwar: A History of Europe Since 1945*. New York: Penguin Books.

Junker, B., M. Buchecker, and U. Muller-Boker. 2007. "Objectives of Public Participation: Which Actors Should Be Involved in the Decision Making for River Restorations?" *Water Resources Research* 43: 10.

Keeney, R. L., and D. von Winterfeldt. 2001. "Appraising the Precautionary Principle: A Decision Analysis Perspective." *Journal of Risk Research* 4 (2): 191–202.

Kiker, Clyde F., J. Walter Milon, and Alan W. Hodges. 2001. "Adaptive Learning for Science-Based Policy: The Everglades Restoration." *Ecological Economics* 37 (3): 403–16.

Knight, F. H. 1921. *Risk, Uncertainty and Profit*. Chicago: University of Chicago Press.
Knight, Jack, and James Johnson. 1994. "Aggregation and Deliberation: On the Possibility of Democratic Legitimacy." *Political Theory* 22 (2): 277–96.
Kraft, Michael E., and Norman J. Vig. 2010. "Environmental Policy over Four Decades: Achievements and New Directions." In *Environmental Policy: New Directions for the Twenty-First Century*, 7th ed., edited by Norman J. Vig and Michael E. Kraft, 1–27. Washington, DC: CQ Press.
Kranzer, Bonnie. 2003. "Everglades Restoration: Interactions of Population and Environment." *Population and Environment* 24 (6): 455–84.
Krimsky, S., and D. Golding, eds. 1992. *Social Theories of Risk*. Westport, CT: Praeger.
Kweit, M. G., and R. W. Kweit. 1987. "The Politics of Policy Analysis." In *Citizen Participation in Public Decision Making*, edited by J. DeSario and S. Langton. New York: Greenwood Press.
Lackey, R. 2007. "Science, Scientists, and Policy Advocacy." *Conservation Biology* 21 (1): 12–17.
Lancaster, K. 1966. "A New Approach to Consumer Theory." *Journal of Political Economy* 74: 132–57.
Lee, K. N. 1989. "The Columbia River Basin: Experimenting with Sustainability." *Environment* 31 (6): 6–11, 30–33.
———. 1993. *Compass and Gyroscope: Integrating Science and Politics for the Environment*. Washington, DC: Island Press.
———. 2001. "Appraising Adaptive Management." In *Biological Diversity: Balancing Interests through Adaptive Collaborative Management*, edited by Louise E. Buck, Charles C. Geisler, John Schelhas, and Eva Wallenberg, 3–25. Washington, DC: CRC Press.
Lichtenstein, S., P. Slovic, B. Fischhoff, M. Layman, and B. Combs. 1978. "Judged Frequency of Lethal Events." *Journal of Experimental Psychology: Human Learning and Memory* (4): 551–78.
Luce, R. D., and J. W. Tukey. 1964. "Simultaneous Conjoint Measurement: A New Type of Fundamental Measurement." *Journal of Mathematical Psychology* 1 (2): 1–27.
Lynam, Timothy, Wil de Jong, Douglas Sheil, Trikurnianti Kusumanto, and Kirsten Evans. 2007. "A Review of Tools for Incorporating Community Knowledge, Preferences, and Values into Decision Making in Natural Resources Management." *Ecology and Society* 12 (1): 5. http://www.ecologyandsociety.org/vol12/iss1/art5/.
Machlis, G. E., and E. Rosa. 1990. "Desired Risk: Broadening the Social Implication of Risk Framework." *Risk Analysis* 10: 161–68.
Manson, Neil. 2002. "Formulating the Precautionary Principle." *Environmental Ethics* 24 (3): 263–74.
Marsella, A. J. 1998. "Toward a 'Global-Community Psychology': Meeting the Needs of a Changing World." *American Psychologist* 53: 1282–91.
Martin, I. M., and T. A. Steelman. 2004. "Using Multiple Methods to Understand Agency Values and Objectives: Lessons from Public Lands Management." *Policy Sciences* 37: 37–69.

Mascarenhas, Michael, and Rik Scarce. 2004. "'The Intention was Good': Legitimacy, Consensus-based Decision Making, and the Case of Forest Planning in British Columbia, Canada." *Society & Natural Resources* 17 (1): 17–38.

Mason, R. O., and I. I. Mitroff. 1973. "A Program for Research on Management Information Systems." *Management Science* 19 (5): 475–87.

McCabe, J. Terrence. 2003. "Sustainability and Livelihood Diversification among the Maasai of Northern Tanzania." *Human Organization* 62 (2): 100–111.

McKelvey, K. S., and J. D. Johnson. 1992. "Historical Perspectives on Forests of the Sierra Nevada and the Transverse Ranges of Southern California: Forests at the Turn of the Century." In *The California Spotted Owl: A Technical Assessment of its Current Status*, 225–46. PSW-GTR-133. Albany, CA: USDA Forest Service Pacific Southwest Research Station. http://www.treesearch.fs.fed.us/pubs/3536.

McKeown, B., and D. Thomas. 1988. *Q methodology*. Newbury Park: Sage.

McLain, R. J., and R. G. Lee. 1996. "Adaptive Management: Promises and Pitfalls." *Environmental Management* 20: 437–48.

McRae, Duncan Jr., and Dale Whittington. 1997. *Expert Advice for Policy Choice: Analysis and Discourse*. Washington, DC: Georgetown University Press.

Mealey, S. P., J. W. Thomas, H. J. Salwasser, R. E. Stewart, P. J. Balint, and P. W. Adams. 2005. "Precaution in the American Endangered Species Act as a Precursor to Environmental Decline: The Case of the Northwest Forest Plan." In *Biodiversity and the Precautionary Principle: Risk and Uncertainty in Conservation and Sustainable Use*, edited by R. Cooney and B. Dickson, 189–204. London: Earthscan.

Meretsky, V. J., D. L. Wegner, and L. E. Stevens. 2000. "Balancing Endangered Species and Ecosystems: A Case Study of Adaptive Management in Grand Canyon." *Environmental Management* 25 (6): 579–86.

Meyer, John M. 1997. "Gifford Pinchot, John Muir, and the Boundaries of Politics in American Thought." *Polity* 30 (2): 267–84.

Miller, George A. 1956. "The Magical Number Seven, Plus or Minus Two." *Psychological Review* 63 (2): 81–97.

Mitchell-Banks, Paul. 2006. "Participatory Process as a Tool to Resolve Conflict." *Schweizerische Zeitschrift fur Forstwesen* 157 (10): 471–76.

Mitroff, Ian A., and Francisco Sagasti. 1973. "Epistemology as General Systems Theory: An Approach to the Design of Complex Decision-making Experiments." *Philosophy of the Social Sciences* 3: 117–34.

Moir, W. H., and W. M. Block. 2001. "Adaptive Management on Public Lands in the United States: Commitment or Rhetoric?" *Environmental Management* 28 (2): 141–48.

Muir, J. 1909. *Our National Parks*. Boston: Houghton Mifflin.

Nash, Roderick E. 1989. *The Rights of Nature: A History of Environmental Ethics*. Madison: University of Wisconsin Press.

National Research Council. 1989. *Improving Risk Communication*. Washington, DC: National Academy Press.

———. 1996. *Understanding Risk: Informing Decisions in a Democratic Society*. Washington, DC: National Academy Press.
———. 2006. *Progress toward Restoring the Everglades: The First Biennial Review, 2006*. Washington, DC: National Academy Press.
NEJAC. 2000. The Model Plan for Public Participation. Washington, DC: Office of Environmental Justice, US EPA. http://www.edf.org/documents/2814_modelbk.pdf.
Nelson, Robert H. 1999. "The Religion of Forestry: Scientific Management." *Journal of Forestry* 97 (11): 4–8.
Neumann, R. 1998. *Imposing Wilderness: Struggles over Livelihood and Nature Preservation in Africa*. Berkeley: University of California Press.
Nordhaus, William. 2007. "Critical Assumptions in the Stern Review on Climate Change." *Science* 317 (5835): 201–2.
O'Brien, E. 2003. "Human Values and Their Importance to the Development of Forestry Policy in Britain: A Literature Review." *Forestry* 76: 3–17.
Olsson, P., C. Folke, and F. Berkes. 2004. "Adaptive Comanagement for Building Resilience in Social-Ecological Systems." *Environmental Management* 34 (1): 75–90.
O'Reilly, C. A., J. A. Chatman, and D. F. Caldwell. 1991. "People and Organizational Culture: A Profile Comparison Approach to Assessing Person-Organization Fit." *Academy of Management Journal* 34: 487–516.
O'Riordan, T., and A. Jordan. 1995. "The Precautionary Principle in Contemporary Environmental Politics." *Environmental Values* 4: 191–212.
Ostergren, David M., Kimberly A. Lowe, Jesse B. Abrams, and Elizabeth J. Ruther. 2006. "Public Perceptions of Forest Management in North Central Arizona: The Paradox of Demanding More Involvement but Allowing Limits to Legal Action." *Journal of Forestry* 104 (7): 375–82.
Pahl-Wostl, C., J. Sendzimir, P. Jeffrey, J. Aerts, G. Berkamp, and K. Cross. 2007. "Managing Change toward Adaptive Water Management through Social Learning." *Ecology and Society* 12 (2): 30. http://www.ecologyandsociety.org/vol12/iss2/art30/main.html.
Paris, David C., and James F. Reynolds. 1983. *The Logic of Policy Inquiry*. New York: Longman.
Pateman, C. 1970. *Participation and Democratic Theory*. New York: Cambridge University Press.
Patton, Carl V., and David S. Sawicki. 1993. *Basic Methods of Policy Analysis and Planning*, 2nd ed. Englewood Cliffs, NJ: Prentice Hall.
Pellizzoni, Luigi. 2003. "Uncertainty and Participatory Democracy." *Environmental Values* 12 (2): 195–224.
Perrow, C. 1984. *Normal Accidents: Living with High Risk Technologies*. New York: Basic Books.
Peterson, Martin. 2002. "The Limits of Catastrophe Aversion." *Risk Analysis* 22 (3): 527–38.

Petts, Judith, and Catherine Brooks. 2006. "Expert Conceptualisations of the Role of Lay Knowledge in Environmental Decisionmaking: Challenges for Deliberative Democracy." *Environment and Planning A* 38 (6): 1045–59.

Pielke, Roger Jr. 2002. "Better Safe Than Sorry." *Nature* 419 (6906): 433–4.

Pill, J. 1971. "The Delphi Method: Substance, Context, a Critique and an Annotated Bibliography." *Socio-Economic Planning Science* 5: 57–71.

Pinchot, G. 1947. *Breaking New Ground*. Washington, DC: Island Press.

Pinkerton, E. 1999. "Factors Overcoming Barriers to Implementing Co-management in British Columbia Salmon Fisheries." *Conservation Ecology* 3 (2): 2. http://www.ecologyandsociety.org/vol3/iss2/art2/.

Pittman, Craig. 2007. "Scientists Opposed Glades Decision." *St. Petersburg Times*, July 31.

Poole, Robert M. 2006. "Heartbreak in the Serengeti." *National Geographic*, February.

Pretty, J. 2006. "Social Capital and the Collective Management of Resources." In *State of the Planet 2006–2007*, edited by D. Kennedy, 142–48. Washington, DC: Island Press.

Putnam, R. D., R. Leonardi, and R. Y. Nanetti. 1993. *Making Democracy Work: Civic Traditions in Modern Italy*. Princeton: Princeton University Press.

Raffensperger, C., and J. Tickner. 1999. *Protecting Public Health and the Environment: Implementing the Precautionary Principle*. Washington, DC: Island Press.

Raiffa, H. 1968. *Decision Analysis*. Reading, MA: Addison-Wesley.

Reddy, V. S., R. J. Bush, and R. Roudik. 1995. A Market-oriented Approach to Maximizing Product Benefits: Cases in U.S. Forest Product Industries. Helsinki: University of Helsinki Department of Forest Economics, Publication no. 4. http://www.srs.fs.usda.gov/pubs/ja/ja_reddy001.pdf.

Reed, Mark S. 2008. "Stakeholder Participation for Environmental Management: A Literature Review." *Biological Conservation* 141 (10): 2417–31.

Reed, Mark S., Anil Graves, Norman Dandy, Helena Posthumus, Klaus Hubacek, Joe Morris, Christina Prell, Claire H. Quinn, and Lindsay C. Stringer. 2009. "Who's In and Why? A Typology of Stakeholder Analysis Methods for Natural Resource Management." *Journal of Environmental Management* 90 (5): 1933–49.

Renn, O. 1995. "Style of Using Scientific Expertise: A Comparative Framework." *Science and Public Policy* 22: 147–56.

———. 2006. "Participatory Processes for Designing Environmental Policies." *Land Use Policy* 23 (1): 34–43.

Renn, Ortwin, Thomas Webler, and Peter Wiedemann, eds. 1995. *Fairness and Competence in Citizen Participation: Evaluating Models for Environmental Discourse*. Dordrecht, Germany: Kluwer Academic Publishers.

Rescher, N. 1969. *Introduction to Value Theory*. Englewood Cliffs, NJ: Prentice-Hall.

Riker, William H. 1982. *Liberalism against Populism*. San Francisco: W. H. Freeman.

Rittel, H. W. J., and M. M. Webber. 1973. "Dilemmas in a General Theory of Planning." *Policy Sciences* 4: 155–169.

Rosa, E. A., and D. L. J. Clark. 1999. "Historical Routes to Technological Gridlock: Nuclear Technology as Prototypical Vehicle." *Research in Social Problems and Public Policy* 7:21–57.

Rosa, E., A. Mazur, and T. Dietz. 1987. "Sociological Analysis of Risk Impacts Associated with the Siting of a High Level Nuclear Waste Repository: The Case of Hanford." Paper presented at the Workshop on Assessing Social and Economic Effects of Perceived Risk, Seattle: Battelle Human Affairs Research Centers.

Rowe, Gene, and Lynn J. Frewer. 2000. "Public Participation Methods: A Framework for Evaluation." *Science, Technology & Human Values* 25 (1): 3–29.

Rowe, Gene, Tom Horlick-Jones, John Walls, Wouter Poortinga, and Nick F. Pidgeon. 2008. "Analysis of a Normative Framework for Evaluating Public Engagement Exercises: Reliability, Validity and Limitations." *Public Understanding of Science* 17 (4): 419–41.

Ryan, M., and S. Farrar. 2000. "Using Conjoint Analysis to Elicit Preferences for Health Care." *British Medical Journal* 320: 1530–33.

Saaty, T. L. 1980. *Analytic Hierarchy Process*. New York: McGraw Hill.

Salwasser, H. 2002. "Navigating Through the Wicked Messiness of Natural Resource Problems: Roles for Science, Coping Strategies, and Decision Analysis." Paper presented at the Sierra Science Summit, Kings Beach, CA.

———. 2004. "Confronting the Implications of Wicked Problems: Changes Needed in Sierra Nevada National Forest Planning and Problem Solving." In *Proceedings of the Sierra Nevada Science Symposium: Science for Management and Conservation*. General technical report PSW-GTR-193, 7-22. USDA Forest Service, Kings Beach, CA.

Sample, V. A. 1993. "A Framework for Public-Participation in Natural-Resource Decision-making." *Journal of Forestry* 91 (7): 22.

Sandman, P. M., D. B. Sachsman, M. P. Greenberg, and M. Gotchfeld. 1987. *Environmental Risk and the Press*. New Brunswick, NJ: Transaction Books.

Schwartz, S. H., and W. Bilsky. 1987. "Toward a Universal Psychological Structure of Human Values." *Journal of Personality and Social Psychology* 53: 550–62.

Selin, S., and D. Chavez. 1995. "Developing a Collaborative Model for Environmental Planning and Management." *Environmental Management* 19 (2): 189–95.

Selin, S., C. Pierskalla, D. Smaldone, and K. Robinson. 2007. "Social Learning and Building Trust Through a Participatory Design for Natural Resource Planning." *Journal of Forestry* 105 (8): 421–25.

Selin, S. W., M. A. Schuett, and D. S. Carr. 1997. "Has Collaborative Planning Taken Root in the National Forests?" *Journal of Forestry* 95 (5): 25–28.

———. 2000. "Modeling Stakeholder Perceptions of Collaborative Initiative Effectiveness." *Society & Natural Resources* 13: 735–45.

Senge, P. M. 1990. *The Fifth Discipline: The Art and Practice of the Learning Organization*. New York: Doubleday.

Shamir, M., and J. Shamir. 1995. "Competing Values in Public Opinion: A Conjoint Analysis." *Political Behavior* 17 (1): 107–33.

Shannon, M. A. 1992. "Foresters as Strategic Thinkers, Facilitators, and Citizens." *Journal of Forestry* 90 (10): 24–27.

Sheikh, Pervaze A., and Nicole T. Carter. 2008. *South Florida Ecosystem Restoration and the Comprehensive Everglades Restoration Plan*. Washington, DC: Congressional Research Service. http://www.nationalaglawcenter.org/assets/crs/RS20702.pdf.

Shindler, Bruce A., Mark Brunson, and George H. Stankey. 2002. *Social Acceptability of Forest Conditions and Management Practices: A Problem Analysis*. PNW-GTR-537. Washington, DC: USDA Forest Service. http://www.fs.fed.us/pnw/pubs/pnw_gtr537.pdf.

Shindler, B., and K. A. Cheek. 1999. "Integrating Citizens in Adaptive Management: A Propositional Analysis." *Conservation Ecology* 3 (1): 9. http://www.ecologyandsociety.org/vol3/iss1/art9/.

Shocker, A. D., and V. Srinivasan. 1977. "A Consumer Based Methodology for the Identification of New Product Ideas." *Management Science* 20: 921–37.

Simon, Herbert A. 1957. *Models of Man: Social and Rational*. New York: Wiley.

———. 1960. *The New Science of Management Decision*. New York: Harper and Row.

———. 1997. *Models of Bounded Rationality. Vol. 3: Empirically Grounded Economic Reason*. Cambridge, MA: MIT Press.

Skinner, Carl N., and Chi-Ru Chang. 1996. "Fire Regimes, Past and Present." In *Sierra Nevada Ecosystem Project: Final Report to Congress, vol. 2, Assessments and Scientific Basis for Management Options*. Davis: University of California Centers for Water and Wildland Resources. http://gis.fs.fed.us/psw/programs/ecology_of_western_forests/publications/publications/1996-01-SkinnerChang.pdf.

Sklar, F. H., H. C. Fitz, Y. Wu, R. Van Zee, and C. McVoy. 2001. "The Design of Ecological Restoration Models for Everglades Restoration." *Ecological Economics* 37 (3): 379–401.

Slovic, P. 1987. "Perception of Risk." *Science* 236: 280–86.

———. 2000. *The Perception of Risk*. Sterling, VA: Earthscan.

Smith, C. L., J. Gilden, B. S. Steel, and K. Mrakovcich. 1998. "Sailing the Shoals of Adaptive Management: The Case of Salmon in the Pacific Northwest." *Environmental Management* 22 (5): 671–81.

Smith, F. 1997. *Environmental Sustainability*. Boca Raton, FL: St. Lucie Press.

South Florida Water Management District and US Army Corps of Engineers. 2003. CERP Guidance Memorandum, CGM Number-Revision 029.00, November 19. http://www.cerpzone.org/documents/cgm/cgm_029.00.pdf.

Stankey, G. H., B. T. Bormann, C. Ryan, B. Shindler, V. Sturtevant, R. N. Clark, and C. Philpot. 2003. "Adaptive management and the Northwest Forest Plan: Rhetoric and Reality." *Journal of Forestry* 101 (1): 40–46.

Starr, C. 1969. "Social Benefit Versus Technological Risk." *Science* 165: 1232–38.

———. 2003. "The Precautionary Principle Versus Risk Analysis." *Risk Analysis* 23 (1): 1–3.

Steelman, T. A. 2001. "Elite and Participatory Policymaking: Finding Balance in a Case of National Forest Planning." *Policy Studies Journal* 29 (1): 71–89.

Steelman, T. A., and L. A. Maguire. 1999. "Understanding Participant Perspectives: Q-methodology in National Forest Management." *Journal of Policy Analysis and Management* 18: 361–88.

Steen, H. K., ed. 1992. *The Origins of the National Forests: A Centennial Symposium*. Durham, NC: Forest History Society.

Stephenson, W. 1935. "Technique of Factor Analysis." *Nature* 136: 297.

——. 1953. *The Study of Behavior: Q-technique and Its Methodology*. Chicago: University of Chicago Press.

——. 1978. "Concourse Theory of Communication." *Communication* 3: 21–40.

Stern, Nicholas. 2007. *The Economics of Climate Change: The Stern Review*. New York: Cambridge University Press.

Stern, Nicholas, and Chris Taylor. 2007. "Climate Change: Risks, Ethics, and the Stern Review." *Science* 317 (5835): 203–4.

Stewart, S. I., V. C. Radeloff, R. B. Hammer, and T. J. Hawbaker. 2007. "Defining the Wildland-Urban Interface." *Journal of Forestry* 105 (4): 201–7.

Stirling, A. 1999. *On Science and Precaution in the Management of Technological Risk, vol.1, A Synthesis of Case Studies*. Seville, Spain: Institute for Prospective Technological Studies. ftp://ftp.jrc.es/pub/EURdoc/eur19056en.pdf.

——. 2006. "Analysis, Participation and Power: Justification and Closure in Participatory Multi-criteria Analysis." *Land Use Policy* 23 (1): 95–107.

——. 2008. "'Opening Up' and 'Closing Down'—Power, Participation, and Pluralism in the Social Appraisal of Technology." *Science Technology & Human Values* 33 (2): 262–94.

——. Stubbs, M., and M. Lemon. 2001. "Learning to Network and Networking to Learn: Facilitating the Process of Adaptive Management in a Local Response to the UK's National Air Quality Strategy." *Environmental Management* 27 (3): 321–34.

Sunstein, Cass. 2003. "Beyond the Precautionary Principle." University of Chicago Law & Economics, Olin Working Paper No. 149; University of Chicago, Public Law Working Paper No. 38. Chicago: University of Chicago Law School. http://papers.ssrn.com/sol3/papers.cfm?abstract_id=307098.

Swedlow, B. 2002. "Toward Cultural Analysis in Policy Analysis: Picking Up Where Aaron Wildavsky Left Off." *Journal of Comparative Policy Analysis: Research and Practice* 4: 267–85.

Theiss-Morse, E. 1993. "Conceptualizations of Good Citizenship and Political Participation." *Political Behavior* 15 (4): 355–80.

Thomas, John Clayton. 1995. *Public Participation in Public Decisions*. San Francisco: Jossey-Bass.

Thomas, J. W. 2003. *Application of the Northwest Forest Plan in National Forests in California*. Sacramento, CA: USDA Forest Service Pacific Southwest Region.

Thompson, W. 1935. "On Complete Families of Correlation Coefficients and Their Tendency to Zero Tetrad-differences." *British Journal of Psychology* 26: 63–92.

Thoreau, H. D. (1849, 1854) 1992. *Walden, and Resistance to Civil Government*. 2nd ed. New York: Norton.

Tickner, J. 1999. "A Map Toward Precautionary Decision-making." In *Protecting Public Health and the Environment: Implementing the Precautionary Principle*, edited by C. Raffensperger and J. Tickner. Washington, DC: Island Press.

Tippett, Joanne, John F. Handley, and Joe Ravetz. 2007. "Meeting the Challenges of Sustainable Development—A Conceptual Appraisal of a New Methodology for Participatory Ecological Planning." *Progress in Planning* 67 (1): 9–98.

Tucker, G. A., and J. Treweek. 2005. "The Precautionary Principle in Impact Assessment: An International Review." In *Biodiversity and the Precautionary Principle: Risk and Uncertainty in Conservation and Sustainable Use*, edited by R. Cooney and B. Dickson, 73–93. London: Earthscan.

Tversky, A., and D. Kahneman. 1982. "Availability: A Heuristic for Judging Frequency and Probability." In *Judgment under Uncertainty: Heuristics and Biases*, edited by D. Kahneman, P. Slovic, and A. Tversky, 163–78. Cambridge: Cambridge University Press.

United Nations. 1992. *Rio Declaration on Environment and Development*. Rio de Janeiro: United Nations. http://www.unep.org/Documents.Multilingual/Default.asp?DocumentID=78&ArticleID=1163.

United Nations Environment Programme. 2008. "Ngorongoro Conservation Area, Tanzania." In *Encyclopedia of Earth*, edited by Cutler J. Cleveland. Washington, DC: Environmental Information Coalition, National Council for Science and the Environment. http://www.eoearth.org/article/Ngorongoro_Conservation_Area,_Tanzania.

U.S. Government Accountability Office. 2007. *South Florida Ecosystem: Restoration is Moving Forward but is Facing Significant Delays, Implementation Challenges, and Rising Costs*. Report to the Committee on Transportation and Infrastructure, House of Representatives. GAO-07-520. May. http://www.gao.gov/new.items/d07520.pdf.

USDA Economic Research Service. 2007. *Background Statistics: U.S. Citrus Market*. http://www.ers.usda.gov/News/citruscoverage.htm.

USDA Forest Service. 1993. *Principal Laws Relating to Forest Service Activities*. Washington, DC: US Government Printing Office.

———. 1995. Chapter 7: "Adaptive Management Area." In *Gifford Pinchot National Forest Plan (Amendment 11)*. Vancouver, WA: Gifford Pinchot National Forest. http://www.fs.fed.us/gpnf/04projects/mgtdir/downloads/amend11/c7up4.pdf.

———. 1997. *Integrating Science and Decision Making: Guidelines for Collaboration among Managers and Researchers in the Forest Service*. FS-608. Washington, DC: USDA Forest Service.

———. 1998. "The Olympic Adaptive Management Area." Olympia, WA: Olympic National Forest. http://www.fs.fed.us/r6/olympic/ecomgt/nwfp/adaptman.htm.

———. 2001a. *Sierra Nevada Forest Plan Amendment Final Environmental Impact Statement*. Sacramento: USDA Forest Service.

———. 2001b. *Sierra Nevada Forest Plan Amendment Record of Decision*. Sacramento: USDA Forest Service. http://www.fs.fed.us/r5/snfpa/library/archives/rod/rod.pdf.

———. 2003. *Sierra Nevada Forest Plan Amendment Management Review and Recommendations*. Sacramento: USDA Forest Service Pacific Southwest Region. http://www.fs.fed.us/r5/snfpa/review/review-report/index.html.

van der Brugge, R., and R. van Raak. 2007. "Facing the Adaptive Management Challenge: Insights from Transition Management." *Ecology and Society* 12 (2): 33. http://www.ecologyandsociety.org/vol12/iss2/art33/.

Verner, J., K. S. McKelvey, B. Noon, R. Gutierrez, G. Gould Jr., and T. W. Beck, eds. 1992. *The California Spotted Owl: A Technical Assessment*. PSW-GTR 133. Albany, CA: USDA Forest Service Pacific Southwest Research Station. http://www.fs.fed.us/psw/publications/documents/psw_gtr133/psw_gtr133_fm.pdf.

Vickers, G. 1965. *The Art of Judgement*. New York: Basic Books.

Volkman, J. M., and W. E. McConnaha. 1993. "Through a Glass, Darkly: Columbia River Salmon, the Endangered Species Act and Adaptive Management." *Environmental Law* 23: 1249–72.

von Moltke, K. 1988. "The Vorsorgeprinzip in West German Environmental Policy." In *Royal Commission on Environmental Pollution, 12th report*, 57–70. London: HMSO. http://www.rcep.org.uk/reports/12-bpeo/1988-12bpeo.pdf.

Walker, Robert, and William D. Solecki. 2001. "South Florida: The Reality of Change and the Prospects for Sustainability." *Ecological Economics* 37 (3): 333–37.

Walters, C. 1997. "Challenges in Adaptive Management of Riparian and Coastal Ecosystems." *Conservation Ecology* 1 (2): 1. http://www.ecologyandsociety.org/vol1/iss2/art1/.

Walters, C. J., L. H. Gunderson, and C. S. Holling. 1992. "Experimental Policies for Water Management in the Everglades." *Ecological Applications* 2 (2): 189–202.

Walters, Lawrence C., James Aydelotte, and Jessica Miller. 2000. "Putting More Public in Policy Analysis." *Public Administration Review* 60 (4): 349–59.

Wanless, Harold R., Randall W. Parkinson, and Lenore P. Tedesco. 1994. "Sea Level Control on Stability of Everglades Wetlands." In *Everglades: The Ecosystem and Its Restoration*, edited by Steven M. Davis and John C. Ogden, 199–222. Boca Raton, FL: St. Lucie Press.

Watzlawick, P., J. Weakland, and R. Fisch. 1974. *Change: Principles of Problem Formation and Problem Resolution*. New York: Norton.

Webler, Thomas. 1995. "'Right' Discourse in Citizen Participation: An Evaluative Yardstick." In *Fairness and Competence in Citizen Participation: Evaluating Models of Environmental Discourse*, edited by O. Renn, T. Webler, and P. Wiedemann. Dordrecht: Kluwer Academic Publishers.

Webler, Thomas, and Seth Tuler. 2006. "Four Perspectives on Public Participation Process in Environmental Assessment and Decision Making: Combined Results from 10 Case Studies." *Policy Studies Journal* 34 (4): 699–722.

Wildavsky, A. 1979. *The Art and Craft of Policy Analysis*. London: Macmillan.
Wilhere, G. F. 2002. "Adaptive Management in Habitat Conservation Plans." *Conservation Biology* 16 (1): 20–29.
Wondolleck, J. M. 1988. *Public Lands Conflict and Resolution: Managing National Forest Disputes*. New York: Plenum Press.
Wondolleck, Julia M., and Steven L. Yaffee. 2000. *Making Collaboration Work*. Washington, DC: Island Press.
World Bank. 2006. *Attracting Investment in Tourism, Tanzania's Investor Outreach Program*. Washington, DC: World Bank Group Multilateral Investment Guarantee Agency. http://www.ifc.org/ifcext/fias.nsf/AttachmentsByTitle/MIGA_Tanzania_Tourism/$FILE/tanzaniaweb.pdf.
Yohe, Gary W., Richard S. J. Tol, Richard G. Richels, and Geoffrey J. Blanford. 2008. *Copenhagen Consensus Challenge Paper: Global Warming*. Copenhagen: Copenhagen Consensus Center. http://www.scribd.com/doc/7988846/Copenhagen-Consensus-2008-Yohe#fullscreen:on.
Zinkhan, F. C., T. P. Holmes, and D. E. Mercer. 1997. "Conjoint Analysis: A Preference-based Approach for the Accounting of Multiple Benefits in Southern Forest Management." *Southern Journal of Applied Forestry* 21 (4): 180–86.

AUTHORS

PETER J. BALINT is associate professor at George Mason University with a joint appointment in the Department of Public and International Affairs and the Department of Environmental Science and Policy. His research focuses on natural resource management, community-based conservation, and the life histories of environmental agencies.

RONALD E. STEWART was a career employee of the US Forest Service. He served as regional forester in the Pacific Southwest Region from 1990 to 1994 and was involved at the beginning of the Sierra Nevada Forest Plan Amendment process. He retired from the agency in 1999 with the rank of deputy chief. He then joined the Department of Environmental Science and Policy at George Mason University as visiting associate professor. He is now retired and living in Pennsylvania.

ANAND DESAI is professor in the John Glenn School of Public Affairs at The Ohio State University. His research interests include measurement of performance and evaluation of the provision of public services. In particular, he has worked on methods for measuring effectiveness and efficiency in the public sector and the use of statistical and operations research models for public policy analysis.

LAWRENCE C. WALTERS is Stewart Grow Professor of Public Policy and Management in the Romney Institute at Brigham Young University. His research focuses on the theoretical foundations for public policy analysis and public finance, the analysis of competing value positions, the development of policy arguments, state and local tax policy, and extensions of data envelopment analysis and productivity assessment.

INDEX

Figures/photos/illustrations are indicated by "f" and tables by "t", respectively.

acid rain, Germany's policy on, 67–68
active adaptive management, 81, 133
adaptive, deliberative decision process. *See* decision approach, proposed
adaptive management, 65, 73, 75, 79–102, 104, 117, 126, 133, 164, 166, 212
   active, 80, 81, 89, 100, 133
   barriers to, 95, 96–97, 99–100
   collaborative, 79–82, 84–99
   complexity and uncertainty contexts for, 82–84
   conventional forms of, 80–81, 133
   definition and history, 80–84
   examples of using, 84–99
   Forest Service appraisal of, 97–98
   passive, 80, 89, 100, 133
   precautionary principle conflict with, 83, 98
   social conflicts incorporated into, 81–82
   summary of cases in, 99–102
   value judgments and, 90
Adaptive Management Areas, Northwest Forest Plan's, 95–96
administrative uncertainty, 19, 20
Africa
   CBNRM projects in, 47
   colonization of East, 44–46
Africa Wildlife Foundation, 46
agencies, public, 85, 100–101, 104, 130, 135. *See also* Forest Service
   NEPA process and, 134f
   public distrust of, 159–160, 212–216
aggregation processes, 117, 118f, 120–121, 126, 199
   decision approach, 132, 140
   outcomes, 123–124, 126f
   preference approval voting during, 122–123, 123t
air quality standards, in Great Britain, 84–85
airport debate, Everglades, 41
analysis. *See also* data analysis, preference elicitation
   cost-benefit, 23, 69–73
   economic risk, 23
   formal models for data, 200–204
   risk, 69–72
   scientific, 119
analytical proposal, wicked, 191–200. *See also* decision approach, proposed; preference elicitation, SNFPA
   feasibility, 193–195
   hybrid simulation in, 192–193, 194t
   limitations, 195–196
   technical appendix, 198–200
analytical road map, preference elicitation, 169f
analytic-deliberative processes, 107, 114, 129–148. *See also* decision approach, proposed; deliberative processes
   concept model for categories of, 116–118, 118f
   design requirements, 118–121
   feasible alternative output of, 126
   first stage of, 121–124, 126f
   NRC decision approach as, 138

239

"opening up," 120
principles guiding, 118–119
problem formulation in, 139–140
proposed approach to, 121–127, 126f
summary of participatory approach, 125–127
anthropology, risk assessment from view of, 26–27
appreciative system, Vickers', 21, 22, 22f, 30
approval votes, 122
Arabs, Omani, 43
Arusha Declaration, 46
attributes and descriptions, preference elicitation, 170–171, 171t–173t, 182t, 188, 197t
Audubon Society, 60
Australia, water projects in, 91–93

backcasting, 71
barriers, adaptive management, 95, 96–97, 99–100
biases, risk assessment, 24, 26–27
Big Cypress National Preserve, 36
Bradbury process, 88, 89
Broward, Napoleon, 35–36
Bush, Jeb, 40, 41

California, spotted owls in, 98
Canada, aboriginal rights, 90
cap-and-trade system, 51–55
carbon dioxide emissions. *See* European Union
card-sort exercise, preference elicitation, 173–183, 174f. *See also* data analysis, preference elicitation
data analysis methods overview, 175
data format for, 205
preferences and trade-offs, 174–176
scoring and sorting, 175, 181–183, 182t
cases, wicked, 33–63. *See also* Sierra Nevada; Sierra Nevada Forest Plan Amendment
adaptive management summary of, 99–102
collaborative adaptive management in, 84–99
conclusions, 62–63
conflicting values in, 151–152, 216
consensus lacking in all four, 70–71
decision approach applied to, 142–147
EU emissions, 49–55, 68–71, 146–147, 214, 216
Everglades, 34–43, 70, 93–95, 142, 143–144, 212–213
fuel treatment, 106
NCA, 43–49, 70, 144–146, 215
overview, 33–34
catch-22 situations, policy experimentation, 97, 99–100
categories of uncertainty, 18–20
CBNRM. *See* community-based natural resource management
Central and Southern Florida Project, 36–40
CERP. *See* Comprehensive Everglades Restoration Plan
Chafee, John, 40
Chiles, Lawton, 40, 41
citizen-agency interactions, 100–101
Clean Air Act of 1970, 4
Clean Water Act of 1972, 4
Clinton, Bill, 68, 95
cognitive competence, 115
collaborative adaptive management, 79–82
cases using, 84–99
conditions favoring, 82
Everglades, 93–95
factors in failure of, 94
purpose of, 81
colonial Tanganyika, 44–46
colonization, Maasai impacted by, 44–45
Columbia River Basin, salmon in, 86–87
commons, 14
communication
interagency, 85
learning network, 140
one-way and two-way, 131

community of interest, stakeholder, 15–16
community of place, stakeholder, 15–16
community-based natural resource management (CBNRM), 47
compass and gyroscope metaphor, 86, 94
competence
 fairness and, 110
 normative, 115–116
complexity
 adaptive management in context of, 82–84
 risk assessment, 28
 wicked problems, 15–16, 82
Comprehensive Everglades Restoration Plan (CERP), 39–43, 143
computation, Q-methodology, 202–203
Condorcet winner, 195
conjoint analysis, 175, 176–183, 177t, 198–202. *See also* Q-type analysis
 background, 199–200
 formal models, 201–202
 interpreting, 176–178, 177t
 marketing and business use of, 199–200
 part-worth, 176, 201
 results, 179–183, 182t
 simulation based on, 181–183, 182t
 usefulness and limitations of, 183, 195–196, 198
 utility, 176, 177t, 178, 181
consensus
 cases as lacking, 70–71
 cultural, 26–27
 SNFPA, 156–158, 157t
 social, 215–216
conservation, preservation compared to, 57
constructivism, appreciative system as, 21
continuum of engagement, public participation, 111
controversy. *See* environmental controversies

cost-benefit analysis, precautionary principle compared to, 70, 71
Creative Act of 1891, 58
cultural consensus, 26–27

data analysis, general, 199–204
data analysis, preference elicitation, 141, 176–196
 analytical proposal based on, 191–196
 attributes and rank ordering of six options, 197t
 conjoint analysis, 175, 176–183, 177t, 182t, 194t, 195–196, 198–202
 methods overview, 175
 Q-methodology basis of, 175
 Q-type analysis, 184–191, 186t–187t, 190f, 194t, 196, 202–205
 simulations in, 142, 181–183, 182t, 188–191
data collection
 deliberative processes, 121–124, 126f
 Sierra Nevada survey, 150, 153–155
data format, card-sort exercise, 205
decision approach, proposed, 126f, 130–133. *See also* analytical proposal, wicked; analytic-deliberative processes; processes
 additional components in, 126f, 137
 aggregation phase of, 132, 140
 case application of, 142–147
 deliberation phase of, 126f, 131–132, 140
 discovery phase, 126f, 130–131
 essential components of, 210
 evaluation phase, 126f, 140–141
 learning networks in, 133, 139–141, 139f
 NEPA process and, 126f, 133–138
 preference elicitation and analysis, 141–142
decision criteria, SNFPA, 168–174, 169f, 171t–173t

242 Index

decision problems. *See also* cases, wicked; precautionary principle; wicked problems
   cultural consensus in, 26–27
   determining seriousness of, 8
   dimensions, 9–10, 10t
   easy to wicked, 10t
   Forest Service, 152–153, 153f
   formulating, 12–13, 139–140
   ill-to well-structured, 8–11
   natural resource, 14–16
   precautionary approach, 17
   risk assessment in, 22–30
   scale and, 14–16, 146–147
   scenarios, 10
   traditional approach, 8, 13–14
   uncertainty categories for, 18–20
deliberative processes, 116–117, 118f, 126f
   data collection for, 121–124, 126f
   decision approach phase of, 126f, 131–132, 140
   iterative and interactive stages of, 126f, 132
   objective of, 120
   principles guiding, 119–120
democracy, 86, 105
demographic trends, 215–217
descriptive statistics, on utilities by group, 196t
developing countries, demographic changes in, 214, 215
development
   Central and Southern Florida Project, 36–40
   Everglades repercussions of, 37–39
   NCA repercussions from, 44–46
   Sierra Nevada repercussions of, 57–62
   South Florida, 35–36
   stages in policy, 111–112
dichotomy trap, 72
discovery processes, 116, 118f, 126f
   decision approach phase of, 126f, 130–131
   principles guiding, 118–119
Disston, Hamilton, 35–36

EAA. *See* Everglades Agricultural Area
Earth Summit, in Rio de Janeiro, 214
East Africa, Great Britain colonization of, 44–46
ecology
   complexity in society and, 82
   values integrated with, 101
economic risk analysis, 23
ecosystems, 95
   uncertainty, 16
egalitarian pattern, cultural risk perception, 27
Eight-and-a-Half-Square-Mile Area, 41
Emerson, Ralph Waldo, 57
emissions, carbon dioxide. *See* European Union
Emissions Trading System, EU, 51–55, 146–147, 214, 216
Endangered Species Act, 4, 60
environmental controversies. *See also* cases, wicked; policy, environmental
   Australia water projects, 91-93
   Columbia River Basin salmon, 86–87
   Great Britain air quality standards, 84–85
   Hetch Hetchy River dam, 58
   historical perspective, 3–5
   North America waterfowl harvests, 89–90
   Northwest Forest Plan, 76, 79, 95-98
   Oregon watersheds, 87-89
   social consensus on, 215–216
   spotted owl, 60–62, 74–76, 95–98
   trends contributing to, 212–217
Environmental Defense Fund, 40
environmental impact statement, 125
   decision approach, 132
   draft, 136
   "extraordinary circumstances," 135
   final, 137
   NEPA process, 135, 136
   SNFPA, 74–76

environmental management, 7–31.
　See also managers, public; natural
　　resource management
　categories of uncertainty, 18–20
　living with wicked consequences,
　　13–14
　open and closed ecosystems, 16
　public participation, 104–107
　US environmental movement and, 3
　value neutral uncertainty, 19–20
　values and, 16, 19–22, 22f
Environmental Protection Agency
　(EPA), 104
EPA. See Environmental Protection
　Agency
European Union (EU), emissions
　problem, 49–55, 68–71
　cap-and-trade program implementa-
　　tion, 53
　current situation, 55
　decision approach applied to,
　　146–147
　Emissions Trading System, 51–55,
　　146–147, 214, 216
　global warming framed in, 49–50
　greenhouse gas policy, 68
　policy history, 50–51
　population and GDP, 146
　precautionary principle widely ac-
　　cepted in, 68–69
　scale problems in, 146
　wicked problem characteristics of,
　　49, 53
evaluative processes, 117, 118f, 124
　decision approach phase of, 126f,
　　140–141
Everglades, Florida, 34–43, 70,
　143–144, 216
　airport debate, 41
　Central and Southern Florida Proj-
　　ect, 36–40
　CERP, 39–43, 143
　collaborative adaptive management
　　in, 93–95
　current situation, 41–43
　decision approach applied to,
　　143–144

development repercussions in,
　37–39
environmentalists' internal conflicts
　over, 40
highway construction, 37
hydrological system, 34–35, 38, 41
legislation, 36, 39–40
1990s battles over, 39
political changes and, 40–41
Progressive Era conflicts over, 35–36
recent mitigation efforts, 39–41
scientific uncertainty in, 41
South Florida's dependence on, 142
terrain, 34–35
US Army Corps of Engineers
　responsibility for, 36–37, 39–40,
　212–213
wicked problem characteristics of,
　40–41
Everglades Agricultural Area (EAA),
　36, 37, 39
Everglades National Park, 36, 39
experimentation, policy, 80, 97,
　99–100, 102, 133
experts, public participation and, 161t

factor analysis, 184–185, 186t–187t,
　204
factor loading, 185–186
fairness and competence, Habermasian
　objectives of, 110
fatalistic pattern, cultural risk percep-
　tion, 27
feasibility
　analytical deliberative process out-
　　put, 126
　analytical proposal, 193–195
Flagler, Henry, 35
Florida, Spanish cession of, 35. See also
　Everglades, Florida
Forest Ecosystem Management Assess-
　ment Team, 95
forest management
　environmental policy and, 58–59
　forest regeneration for, 179, 180
Forest Reserve Act, 3–4
forest reserves. See national forests

Forest Service
  adaptive management appraisal by, 97–98
  attitudes towards capacity of, 158–160, 159t
  decentralized administration of, 60
  decision problem of, 152–153, 153f
  first seventy-five years of, 58–59
  legislation, 60–62
  public participation efforts of, 124–125, 130
  public trust in, 212–213
  Sierra Nevada goals of, 74
  SNFPA policy selected by, 75–76
frames
  risk assessment, 28–29
  social amplification framework, 27–28
Framework Convention on Climate Change, 50
fuel treatment
  mechanical techniques for, 170, 179–180, 186–188, 186t
  preference elicitation on, 168–174, 171t–173t
  wicked case on, 106
  wildfire risk addressed by, 74–76

gambling, example of risk perception, 24
game theory, 195
GDP. *See* gross domestic product
geographic scale, 14, 144, 149
geography, natural disaster response, 27–28
Germany, 44
  acid rain policy in, 67–68
global warming
  EU framing of, 49–50
  precautionary principle applied to, 7, 66–67
  scientific uncertainty, 49–50, 66
Government Land Office, 58
Great Britain
  air quality standards in, 84–85
  East African colonization by, 44–46

greenhouse gases, 33–34, 49–55, 68, 146, 214
gross domestic product (GDP), EU, 146

Habermasian objectives, 110
Hetch Hetchy River, dam controversy, 58
hierarchical pattern, cultural risk perception, 27
Homestead Air Force Base, 41
homogeneity, preference, 175
hybrid simulations, 192–193, 194t
hydrology, Everglades, 34–35, 38, 41

ignorance, uncertainty category as, 19
implementation uncertainty, 19, 20
indeterminacy, 19
individualistic pattern, cultural risk perception, 27
industrialization, 213
institutional approach, risk assessment, 25
institutional barriers, adaptive management, 96
institutional champion, 91
instrumental imperative, 111
interagency communication, 85
interest groups, 215
International Decade of the World's Indigenous Peoples, 47
International Union for Conservation of Nature (IUCN), 46

Kelly, George, 22
Kissimmee River, 34
knowledge, value agreement and state of, 10t
Kyoto Protocol, 50, 53, 68

ladder of participation, 111
land management philosophy, 151
landscape scale, 14–16
leadership and definition barriers, adaptive management, 96
learning by doing, 80, 97
learning networks, 85, 114, 129, 138

communication in, 140
decision approach, 133, 139–141, 139f
enhanced, 149–151, 155, 210
iterative cycles in maturing, 153
NEPA process integration with, 126f, 210
preference elicitation and, 211–212
suggested process for, 121–127, 126f
Lee, Kai, 86
legislation. *See also* National Environmental Policy Act; policy, environmental
barriers to natural resource management, 96–97
1890s, 3–4
Everglades, 36, 39–40
first national forests, 58
Forest Service, 60–62
NCA-related, 44–48
1960s, 4
public participation, 104, 196
species viability, 15
locations, attribute description, 170, 172t, 182t, 188, 197t

Maasai, 43–49, 145
majority buy-in, 143
management. *See also* adaptive management; environmental management; managers, public; natural resource management; Sierra Nevada Forest Plan Amendment
community-based, 47
decentralized, 60
forest, 58–59, 179, 180
philosophy of land, 151
risk, 24–25, 28–30
survey questionnaire responses on, 164–166, 165t
wicked problems, 207–208
managers, public, 138–139, 207–212
applied policy context, 190
demographic trends as relevant to, 216–217
personal values of, 125

market-based regulatory policy, 54
marketing, conjoint analysis use in business and, 199–200
Markov processes, 89
maximum principle, 72
mechanical treatment, fuel treatment techniques, 170, 179–180, 186–188, 186t
medical risks, risk perception example of, 29
methodologies. *See also* strategies; *specific strategies*
data analysis overview, 175
outreach, 130–131, 145
Q-, 175, 200, 202, 204
survey, 144
Mexican-American War, 56
minorities, preferences among powerful, 191
models. *See also* theories
analytic-deliberative process concept, 116–118, 118f
conjoint analysis, 201–202
uncertainty, 18
modernization, reflexive, 26
Muir, John, 57–58, 61

National Environmental Policy Act (NEPA), 4, 6, 60, 67, 125
changing attitudes and, 213
feasible alternatives subjected to, 126
public participation mandated by, 104
national forests. *See also* Sierra Nevada
authority over, 3, 4, 57–59
Big Cypress National Preserve, 36
Serengeti National Park, 43, 44, 48
Sierra Nevada, 149, 150f
twentieth-century thought in creation of, 45–46
values, 151–152
Yellowstone National Park as first, 3, 45
National Forest Management Act, 15, 60, 125
National Park Service, 40

National Research Council (NRC), 29, 118, 138, 139f, 196
natural disasters, 27–28, 29
natural resource management, 14–16
  barriers to, 96–97
  community-based, 47
  Forest Service role in, 61
  landscape scale, 14–16
  preservation and conservation of, 57
  time and spatial dimensions of, 15–16
  water, 4, 87–89, 91–93, 94
NCA. *See* Ngorongoro Conservation Area
NEPA. *See* National Environmental Policy Act
NEPA process, 129, 210
  agency use of, 134f
  alternative initial outcomes, 134f, 135
  decision approach as using, 126f, 133–138
  enhanced, 126f, 137, 146–147
  environmental impact statement in, 135, 136
  key addons to, 126f, 137
  learning network integration with, 126f, 210
  requirements, 132, 143, 144, 147, 210
  scale problems addressed by enhanced, 146–147
  US use of, 134f, 135
  worldwide use of, 133
Ngorongoro Conservation Area (NCA), 43–49, 70, 144–146, 215
  assumptions and worldview, 45–46
  current situation, 48–49
  decision approach proposal applied to, 144–146
  development repercussions, 44–46
  disparities in, 144–145
  region of, 43
  resolution efforts, 46–47
Ngorongoro Conservation Area Authority (NCAA), 44–45, 48
Ngorongoro Crater, 43

normative competence, stakeholder, 115–116
normative perspective
  public participation, 110
  social problems from, 20–21
Northwest Forest Plan, 76, 95–98
Northwest Power Planning Council, 87
NRC. *See* National Research Council
Nyerere, Julius, 46, 47

Occupational Safety and Health Act, 67
Omani Arabs, 43
Oregon. *See* Southwestern Oregon, watersheds in
Oregon Water Enhancement Board, 87
Organic Act of 1897, 4
organizational approach, risk assessment, 25
outcomes
  aggregation process, 123–124, 126f
  attribute description, 171, 173t, 182t
  attributes and rank ordering of six options, 197t
  NEPA process alternative initial, 134f, 135
outreach, methods for, 130–131, 145

Pacific Northwest, 60–61. *See also* Columbia River Basin
  Northwest Forest Plan implemented in, 76
Pacific Southwest Region, 59–60, 150f
pair-wise voting, 195
parameter uncertainty, 18
participants, survey questionnaire, 154–156
  preference structure of, 177t
  summary of responses to, 157t
  underlying beliefs of, 211
participatory processes, 107–111. *See also* analytic-deliberative processes; preference elicitation, SNFPA; public participation
  best practices in, 109, 111

decision approach, 126f, 130–133
design of, 107–109
embedding formal processes with, 124–125
summary, 125–127
theory of, 114
part-worth, conjoint analysis, 176, 201
passive adaptive management, 80–81, 133
pastoralism, Maasai, 48–49
people-nature dichotomy, 42–43
perceptions, stakeholder preferences and, 130–131. *See also* risk perception
personal values, 125
pilot testing, Sierra Nevada case, 141, 150–151
Pinchot, Gifford, 57, 58, 61
policy, environmental. *See also* legislation; National Environmental Policy Act
conjoint analysis usefulness for, 183, 198
EU greenhouse gas, 68
experimentation, 80, 97, 99–100, 102, 133
forest management, 58–59
Forest Service SNFPA selected, 75–76
Germany's acid rain, 67–68
history of EU emissions, 50–51
market-based regulatory, 54
preference elicitation conclusions for applicability to, 198
Q-type analysis application to, 190
science issues compared to issues of, 14
stages in developing, 111–112
traditional approach to, 8, 13–14
trends complicating, 212–217
politically induced uncertainty, 19
politics, Everglades and, 40–41
population
EU GDP and, 146
Sierra Nevada, 146
precautionary principle, 17, 65–77, 104, 164. 166

adaptive management conflict with, 83, 98
advantages over current alternatives, 69–71
cost-benefit analysis compared to, 70, 71
emergence and evolution, 67–68
EU acceptance of, 68–69
global warming application of 7, 66–67
overview, 66–67
risk assessment and, 70, 71
SNFPA application of, 73–76
strengths and limitations of, 69–73
survey questionnaire responses on, 166
US skepticism of, 68
variations, 71–73
preferences
approval voting, 121–124, 123t
homogeneity expectation, 175
minority, 191
nuanced, 179–180
stakeholder perceptions and, 130–131
timber harvesting nuanced, 179–180
trade-offs and, 174–176
preference elicitation, learning networks and, 211–212
preference elicitation, SNFPA. *See also* data analysis, preference elicitation
analytical proposal based on, 191–196
analytical road map, 169f
attributes and descriptions, 170–171, 171t–173, 182t, 188, 197t
card-sort exercise, 173–183, 174f, 182t, 205
conclusions, 196–198, 197t
decision criteria, 168–174, 169f, 171t–173t
exercise, 167–168, 169f
hybrid simulations in, 192–193, 194t
policy applicability of, 198
stakeholder preference analysis and, 141–142

248 Index

wildfire use preferences, 177t, 187–188, 189t
preservation, conservation compared to, 57
probabilistic events, 19
problems. *See* decision problems; wicked problems
procedural deficiencies, 4–5
processes. *See also* aggregation processes; analytic-deliberative processes; deliberative processes; NEPA process; participatory processes
   Bradbury, 88, 89
   discovery, 116, 118–119, 118f, 126f, 130–131
   evaluative, 117, 118f, 124, 126f, 140–141
   learning networks suggested, 121–127, 126f
   Markov, 89
   NEPA, 125, 126, 129, 133–138, 144, 147, 210
   SNFPA decision, 160
Progressive Era, 35–36
proposed approaches. *See also* analytical proposal, wicked; decision approach, proposed
public participation, 110–127, 126f
psychology
   conjoint analysis from field of, 199
   risk assessment and, 24–25
public
   agencies distrusted by, 159–160, 212–216
   values of, 103, 127, 151–152, 160
public officials, as stakeholders, 138–139
public participation, 4. *See also* participatory processes; survey questionnaire, Sierra Nevada
   acceptance and resistance to, 124
   applicability, 112
   CERP memorandum on, 143
   questions on role of, 107
   citizen-agency interaction, 100–101
   Columbia River Basin salmon issue, 86–87
   continuum of engagement, 111
   decision process and value of, 160
   design of successful, 109
   embedding formal processes with, 124–125
   environmental management, 104–107
   experts and, 161t
   factors calling for increased, 112, 113t
   Forest Service efforts of, 124–125, 130
   ladder approaches to, 111
   limitations of, 103–104
   majority buy-in and, 143
   as mandated, 104, 196
   outreach methods for, 130–131, 145
   participatory processes for, 107–109
   precedent, 60
   proposed approach for, 110–127, 126f
   risk management, 24–25, 28–30
   risk perception variable in, 105–106
   scale limitations in, 103, 115, 124–125
   science and, 106–107, 138–139
   in scientifically-framed issues, 106–107
   stakeholder distinguished from, 115, 127
   sustained attention, 114
   value diversity due to, 21
Public Participation and Accountability Workgroup, 104

Q-methodology, 175, 200, 202, 204
Q-type analysis, 184–188, 186t–187t, 191, 194t, 196
   data format and factor analysis in, 205
   formal model for, 202–204
   policy application of, 190
   respondent distribution of values, 190f
   results, 184–188

simulations using, 188–191
statistical procedure underlying, 184–188
utility, 192
questionnaire. *See* survey questionnaire, Sierra Nevada

record of decision, 133
Sierra Nevada 2001, 137–138
reflexive modernization, 26
representative democracy, and pure, 105

Rio de Janeiro, Earth Summit in, 214
Rio Declaration, 68
risk
uncertainty and, 7–31
value diversity and, 17
wicked problem contributions by, 138
wildfire, 61–62, 74–76, 106, 168–174, 171t–173t
risk assessment, 9, 24, 25, 70, 72, 76
biases in, 24, 26–27
complexity, 28
economic, 23
precautionary principle and, 70, 71
public participation in, 24–25, 28–30
social sciences perspective on, 22–30
risk management, public participation in, 24–25, 28–30
risk perception
cultural patterns of, 27
gambling example of, 24
medical, 29
public participation with varied, 105–106
social amplification framework, 27–28
risk-risk analysis, 72
Rittel, H. W. J., 12–14, 29–30, 31t
Rockefeller, John D., 35
Roosevelt, Theodore, 57–58

salmon controversy, Columbia River Basin, 86–87

satisficing, 2, 114–115, 207, 209–210
scale
decision problems and, 14–16, 146–147
public participation limited, 103, 115, 124–125
science. *See also* social sciences
democracy and, 86
policy issues compared to those of, 14
public participation and, 106–107, 138–139
scientific analysis, analytic-deliberative process, 119
scientific uncertainty, 19–20, 41, 49–50
adaptive management to reduce, 81
global warming, 49–50, 66
precautionary principle with, 69
values and, 100
wicked problems and, 2–3, 9, 10, 14, 65, 69, 138, 144
scientists, as stakeholders, 138–139
scoring and sorting, card-sort exercise, 175, 181–183, 182t
Seminole Wars, 35
Serengeti National Park, 43, 44, 48
Sierra Club, 57, 60
Sierra Nevada, 5–6, 30, 55–63, 208, 216. *See also* preference elicitation, SNFPA
county borders of, 150f
decision problem, 152–153, 153f
development repercussions for, 57–62
Forest Service goals for, 74
learning networks enhanced approach in, 149–151, 155, 210
Muir's activism in, 57–58, 61
national forests in, 149, 150f
pilot testing in, 141, 150–151
population growth in, 59
record of decision, 137–138
region and settlement of, 55–57
spotted owl controversy in, 60–62, 74–76, 95–98

stakeholder survey, 149–166, 157t, 159t, 161t, 162t–163t, 165t
stakeholders, 73–74
trade-offs, 160–164, 162t–163t, 168, 169f, 170, 174–176
value conflicts in, 151–152
Sierra Nevada Forest Plan Amendment (SNFPA), 60–62, 149–166. *See also* preference elicitation, SNFPA
adaptive management in, 98–99, 164, 166
consensus potential responses on, 156–158, 157t
decision criteria, 168–174, 169f, 171t–173t
decision process and value of public participation, 160
environmental impact statement for, 74–76
Forest Service capacity attitudes, 158–160, 159t
management priorities of, 164–166, 165t
precautionary principle in, 73–76
public participation failure in, 124–125
seven potential compromise options, 189t
summary of responses to statements regarding, 157t
workshop, 153–155, 168–174, 169f, 171t–173t
Simon, Herbert, 2
simulations, 142, 181–183, 182t, 188–193, 211
hybrid, 192–193, 194t
SNFPA. *See* Sierra Nevada Forest Plan Amendment
social amplification framework, risk perception, 27–28
social complexity, ecological and, 82
social consensus, default environmental, 215–216
social problems
adaptive management incorporation of, 81–82

normative conditions creating, 20–21
social sciences
integration of, 28–30
risk assessment viewed by, 22–30
societal trends, 212–216
solutions. *See also* outcomes
living with wicked consequences of, 13–14
satisficing, 2, 114–115, 207, 209–210
South Florida
Central and Southern Florida Project, 36–40
development, 35–36
Everglades dependence in, 142
hydrological systems of, 34–35, 38, 41
people-nature alleged dichotomy in, 42–43
stakeholders, 38–39
terrain, 34–35
South Florida Water Management District, 94
Southwestern Oregon, watersheds in, 87–89
Spain, Florida ceded by, 35
spatial perspective, risk assessment from, 27–28
species, 4, 60
viability, 15–16
spotted owls, 60–62, 74–76, 95–98
stakeholders. *See also* preference elicitation
agencies as not surveying, 130
creative dialogue between, 85
definition of, 130
involvement, 9, 10t
landscape scale and, 15–16
learning networks using preferences of, 127
NCA, 49
normative competence of, 115–116
participatory methods engaging, 110
perceptions and preferences, 130–131
public participation distinguished from, 115, 127

scientists and public officials as, 138–139
Sierra Nevada, 73–74
South Florida, 38–39
survey of Sierra Nevada, 149–166, 157t, 159t, 161t, 162t–163t, 165t
values of, 17, 127, 131
within-group differences, 183
statistics. *See also* simulations
descriptive, 196t
formal models for estimation procedures, 200–204
Q-type analysis, 184–188
statutory and regulatory barriers, adaptive management, 96–97
Stewart, Ron, 208
stochastic uncertainty, 19, 20
strategies. *See also* analytical proposal; decision approach, proposed; precautionary principle; preference elicitation, SNFPA; solutions
adaptive management, 79–102, 133, 164, 166
commonly used, 65
public participation proposed, 110–127, 126f
Vickers' appreciative system, 21, 22, 22f, 30
structural barriers, to adaptive management, 99–100
substantive perspective, public participation, 110–111
sugarcane, 37, 38–39
sulfur dioxide, US cap-and-trade system for, 53
survey methods, 144
survey questionnaire, Sierra Nevada
conclusions, 166
consensus potential responses on, 156–158, 157t
data collection, 150, 153–155
decision process and value of public participation responses, 160
findings from, 155–166, 157t, 159t, 161t, 162t–163t, 165t
follow up, 167–168, 169f

Forest Service capacities, attitudes towards, 158–160, 159t
management philosophy responses on, 164, 165t
management priorities responses, 164–166
overview, 149–151
participants, 154–156, 157t, 177t, 211
structure of, 155–156
trade-offs, 160–164, 162t–163t
values context for, 151–152
sustained attention, public participation, 114
Sydney, catchment area, 92
systemic uncertainty, 18

Tanganyika, 44–47
Tanzania, 5, 7, 33, 43–49, 70, 142, 144–146, 212, 214–216
theories. *See also* models
game, 195
participatory process, 114
traditional decision-making, 8, 13–14
Thomas, J. W., 97
Thoreau, David Henry, 57
timber harvesting, 59, 60, 61, 151
nuanced preferences regarding, 179–180
preference elicitation exercise, 170, 187–188
survey questionnaire responses on, 166
tourism, 47, 48
trade-offs, Sierra Nevada, 160–164, 168
analytical road map to determine, 169f
decision criteria, 170
preferences and, 174–176
survey responses to statements regarding, 160–164, 162t–163t
treatment. *See* fuel treatment
trends
demographic, 215–217
social, 212–216

Umpqua Basin Watershed Council, 87–88
UN. *See* United Nations
UN World Heritage Committee, 39
uncertainty. *See also* scientific uncertainty
  adaptive management in context of, 82–84
  cap-and-trade program, 54–55
  categories of, 18–20
  definition of, 19
  ecosystems, 16
  NEPA process finding of, 134f, 135
  precautionary principle in conditions of, 69
  risk and, 7–31
  types of, 14
  as value neutral, 19–20
UNESCO. *See* United Nations Educational, Scientific, and Cultural Organization
United Nations (UN), 47, 68
United Nations Educational, Scientific, and Cultural Organization (UNESCO), 39, 46
United States (US)
  attitude changes in, 214–215
  environmental movement in, 3
  Florida ceded to, 35
  NEPA process approach in, 134f, 135
  precautionary principle skepticism in, 68
  sulfur dioxide cap-and-trade program, 53
US. *See* United States
US Army Corps of Engineers, 36–37, 39–40, 212–213
US Fish and Wildlife Service, 74, 75, 89, 96–97
US Forest Service. *See* Forest Service
utility, 10, 13, 103, 175–178, 181, 192, 194t–195t, 198, 201, 205

values, 1, 20–22. *See also* preferences; preference elicitation, SNFPA
  adaptive management and, 90
  agreement, 10t
  appreciative system of, 21, 22, 22f, 30
  cases' conflicting, 151–152, 216
  changing, 131
  definition of, 20
  disparity in, 144–145
  diversity of, 17, 21
  ecology integration with, 101
  identifying masked, 191
  manager's personal, 125
  on national forests, 151–152
  overlapping, 186t–187t, 190
  preferences and, 127
  public, 103, 127, 151–152, 160
  Q-type analysis respondent distribution of, 190f
  social problems as created by, 20–21
  stakeholder, 17, 127, 131
  uncertainty as neutral in, 19–20
  wicked problems and, 2–3, 5, 9, 10, 14, 63, 69, 138, 144
Vickers, Geoffrey, 21, 22, 22f, 30
voting
  pair-wise, 105
  preference approval, 121–124, 123t

water, 4, 87–89, 91–93, 94. *See also* hydrology, Everglades
waterfowl harvests, 89–90
watersheds, Southwestern Oregon, 87–89
Webber, M., ten propositions of Rittel and, 12–14, 29–30, 31t
wicked problems, 9–11. *See also* cases, wicked; decision problems; managers, public; strategies
  addressing, 30–32, 31t
  analytical proposal for, 191–200, 194t
  appreciative system for, 21, 22, 22f, 30
  belief in perfect solution for, 208–209
  in changing society, 212–216
  common characteristics of, 11–13
  complexity, 15–16

Index   253

concept, 2–3, 5
conditions associated with, 31t
decisions ranging from easy to, 10t
ecological complexity in, 82
identification of, 6, 29–30
living with solution consequences, 13–14
management of, 207–208
Muir and Pinchot contrasting views reflected in, 57
public participation proposed approach for, 110–127, 126f
risk contribution to, 138
Rittel and Webber ten propositions, 12–14, 29–30, 31t
scientific uncertainty and, 2–3, 9, 10, 14, 65, 69, 138, 144
values and, 2–3, 5, 9, 10, 14, 20–22, 22f, 63, 69, 138, 144
wildfires, prescribed, 177t, 187–188, 189t

wildfire risk, 61–62
  fuel treatment to address, 74–76
  preference elicitation on, 168–174, 171t–173t
  public participation and, 106
wildland-urban intermix (WUI), 98, 164, 168, 170, 172t, 174, 179, 180
wildlife
  NCA, 43, 45–46, 48–49
  Sierra Nevada, 55–56
within-group differences, 183
workshops. *See also* preference elicitation, SNFPA
  Australian water project, 91–93
  Everglades, 93–95
  SNFPA, 153–155, 168–174, 169f, 171t–173t
WUI. *See* wildland-urban intermix

Yellowstone National Park, 3, 45
Yosemite Valley, 57–58

Island Press | Board of Directors

DECKER ANSTROM *(Chair)*
Board of Directors
Comcast Corporation

KATIE DOLAN *(Vice-Chair)*
Conservationist

PAMELA B. MURPHY *(Treasurer)*

CAROLYN PEACHEY *(Secretary)*
President
Campbell, Peachey & Associates

STEPHEN BADGER
Board Member
Mars, Inc.

MERLOYD LUDINGTON LAWRENCE
Merloyd Lawrence, Inc.
  and Perseus Books

WILLIAM H. MEADOWS
President
The Wilderness Society

DRUMMOND PIKE
Founder
Tides

ALEXIS G. SANT
Managing Director
Persimmon Tree Capital

CHARLES C. SAVITT
President
Island Press

SUSAN E. SECHLER
President
TransFarm Africa

VICTOR M. SHER, ESQ.
Principal
Sher Leff LLP

PETER R. STEIN
General Partner
LTC Conservation Advisory
  Services
The Lyme Timber Company

DIANA WALL, PH.D.
Director, School of Global
Environmental Sustainability
  and Professor of Biology
Colorado State University

WREN WIRTH
President
Winslow Foundation